Statistics Tutor for
EXPLORING STATISTICS
A Modern Introduction
to Data Analysis
and Inference

2nd Edition

LARRY J. KITCHENS
ANITA NARVARTE KITCHENS
Appalachian State University

Duxbury Press
An Imprint of Brooks/Cole Publishing Company

I(T)P® An International Thomson Publishing Company

Pacific Grove • Albany • Belmont • Bonn • Boston • Cincinnati • Detroit
Johannesburg • London • Madrid • Melbourne • Mexico City • New York
Paris • Singapore • Tokyo • Toronto • Washington

Assistant Editor: *Cindy Mazow*
Editorial Associate: *Rita Jaramillo*
Marketing Team: *Carolyn Crockett*
Production Editor: *Mary Vezilich*
Printing and Binding: *The Maple-Vail Book Manufacturing Group Inc.*

For more information, contact:

BROOKS/COLE PUBLISHING
511 Forest Lodge Road
Pacific Grove, CA 93950
USA

International Thomson Editores
Seneca 53
Col. Polanco
11560 México, D. F., México

International Thomson Publishing Europe
Berkshire House 168-173
High Holborn
London WC1V 7AA
England

International Thomson Publishing Japan
Hirakawacho Kyowa Building, 3F 418
2-2-1 Hirakawacho
Chiyoda-ku, Tokyo 102
Japan

Thomas Nelson Australia
102 Dodds Street
South Melbourne, 3205
Victoria, Australia

International Thomson Publishing Asia
221 Henderson Road
#05-10 Henderson Building
Singapore 0315

Nelson Canada
1120 Birchmount Road
Scarborough, Ontario
Canada M1K 5G4

International Thomson Publishing GmbH
Königswinterer Strasse
53227 Bonn
Germany

Printed in the United States of America

5   4   3   2

ISBN 0-534-26348-8

CONTENTS

Preface

Preface

Many students who take a statistics course feel that they are venturing into a world of the unknown. Many are apprehensive and scared because they really don't know what is ahead and rumor has it that statistics is "hard." Well, it is true. Statistics can be made hard and unforgiving if you, the learner, study and think in ways that are unproductive. For example, procrastination in studying can "kill" any chance of success, but many students don't realize that, they procrastinate and consequently have a terrible time, and then blame it on some kind of lack of ability. On the other hand, statistics can be challenging and interesting if more productive ways of thinking and doing are acquired from the beginning of the course. The key is that you, the student, are in control of your thinking, behavior, and consequently, your learning. The teacher is a facilitator and will help you in many ways, but you control your mind-set which involves an attitude that will be instrumental in your success. This statistical tutor is a guide to show you specific techniques to acquire on your way to success. If you begin on the first day with the "first-day survival kit" in Part A and develop a positive mind-set, then parts B and C will provide more practice and more help in the actual doing of statistics. Part B aids the student with test preparation. It contains a chapter summary, important definitions, a glossary of terms and two practice tests for each chapter. Part C provides the solutions to the odd-numbered problems (only the answers are given in the text book) and is helpful in daily study. This book is your tutor. Along with your professor, your text, your classmates and your own natural ability you can achieve success in statistics.

The Statistics Tutor

<u>Part A</u>

Developing a Mind-Set for Success

1.   Introduction

Welcome to the world of statistics. I say "welcome" because many times students don't feel particularly "welcome" as they enter a new and very different world of study. As you enter the Statistics world–this "different world"—it is important to be comfortable in a feeling that invites you to learn and understand. From experience, with any project you have taken on in the past, you know that the more confident you feel and the more you know about how to approach the project, the more able you are to accomplish your goal of completing the project with a feeling of satisfaction. In a similar way to learning to play an instrument or to play a sport, the more invited you feel and the more techniques you acquire, the easier it is. Oddly enough, your success in becoming more comfortable has more to do with your own view point—your own mind-set than any other factor. You—not your teacher—are in control of your mind-set and consequently your learning. You can become aware of the choices that you have made rather impulsively in the past, and choose to change those that have been unproductive.

This student guide will begin by helping you become more aware of your own mind-set—and its importance for success. Does your attitude and your reactions to what happens in class matter at all? If you are happy or unhappy, frightened or relaxed, does it matter? This guide will help you become aware of the less tangible influences on your success—attitude, reactions, approaches to thinking, and beliefs as they affect results and also give you tangible suggestions for success.   This study guide will:

- discuss the importance of your attitude and reactions.
- suggest how to use the text to your advantage.
- offer you a "first-day survival kit" of study skills which you can begin using today—whether it's your first day of class or not.
- discuss your mind-set for learning statistics including:
     suggestions for studying
     suggestions for thinking
     observations concerning the effect of beliefs.
- give you an approach to this course which will take you out of the classroom where you can see yourself as an employee instead of a student.
- discuss test-taking, including how to study for and take tests successfully.

- help you use self-talk (what you say to yourself) to your advantage.

So, how do you feel? How do you feel about taking statistics? Begin by writing three words which describe your feelings about having to take a statistics class.

_____ , _____ , _____

Now, look at these three words. How positive are they? If the words suggest fear or low energy, begin now to question why you feel that way with the intention of changing these feelings to more positive ones as you begin your work. If these words are negative, find three replacement words. Write three words which describe the feelings which would make your efforts more productive.

_____ , _____ , _____

2.   Your Attitude And Reactions

At this moment in time, what is your attitude toward learning statistics?

_____

_____

_____

One productive attitude is "I want to learn this material more than anything in the world!" Say to yourself, "I want to learn this!" over and over. Why do you want to learn this? Because there is a little determined voice in you that says, "I know I can do this." But there may be another little voice that says, "I've tried this type of course in the past and have always worked hard and had difficulty. Teachers have looked at me with the look of 'there's no hope'." Well there is hope if you go about studying the right way and if you choose a strong positive attitude by reacting positively to everything that happens.

Think about it. You have the choice to react to situations positively or negatively. How do you react to a low test result or how do you react to the everyday up's and down's of class and homework? Throw away your negative reactions, because they only serve to bring you down. Let's look at some examples of positive reactions. Become aware of your reactions and make them all constructively positive.

Here are five sample situations:

1.   You try but can't do the homework.

2.   The teacher goes too fast in class.

3.   The course moves on to a new concept before you're ready.

4.   You feel like the only one who doesn't understand.

5.   You study for tests but the grades are low.

6.   Write one of your own. _____

        _____

Now write your reactions to the above situations:

1.   _____

        _____

2.   _____

        _____

3.   _____

        _____

4.   _____

        _____

5.   _____

        _____

6.   _____

        _____

These reactions are important and can actually energize you if they're positive or stress you if they're negative. You can temper your reactions and keep them constructive no matter how long that first homework assignment is or how low the first test grade is. If you have too much to do and too little time to do it, you will generate the energy to use every minute of your day constructively if you keep your reaction to the work load positive.

Here are some positive reactions to the above situations:

1. I'll come back to this work later today. Maybe I haven't studied enough yet. I'll review my notes and read the book once more. If I still can't get it, I'll ask in class.

2. I'll stop trying to write everything down. Listen carefully, follow the thinking, ask questions, outline the notes and fill in the rest later. If the teacher goes too fast she is assuming that everyone is understanding the material. It's my job to let her know that the material isn't clear. If I can't formulate a question, I still need to respond in some way. Even if I just say "wait a minute, would you go over that again?" If I can approximate where I started to have difficulty I should tell her. Always, stop and ask something.

3. Math-related courses are not intended to be learned in little segments. It is not intended that I fully understand each concept before exploring another one. I'll keep learning as we move on and what follows will help me see the relevance of and understand previous concepts. So, my mind-set should be to learn a concept as fully as I can, then go on to the next as the class moves on. I'll keep reviewing concepts throughout the course, and these will make more sense as I continue with the course.

4. When other people in the class ask questions it always helps me. I should never feel like I'm the only one with a particular question. It helps the instructor teach the class when she knows what her students do and don't understand.

5. Remember, my primary purpose for taking a test is to learn from my mistakes. I should re-do the test and learn any statistical concepts which I haven't as yet mastered. I should also discuss my study skills with my instructor for pointers on studying. I should fine-tune my study techniques, study harder and longer. Then I should expect about a 10 point improvement on the next test. Most often, improvement on tests is gradual. I should allow this gradual improvement without getting frustrated or upset. A gradual upward trend is far better than a downward trend even though the grades may average the same. Teachers sometimes take an upward trend into consideration in formulating a final grade.

6. Write a positive reaction to your situation. _____

_____

Your teacher and text offer an invitation to learn but you actually control every one of your reactions, and thus your energy, your learning, and your sense of comfort.

3. How To Use <u>Exploring Statistics</u>

I've never found that reading a math-related text helps me. Math books confuse me.
~Jerry

<u>Exploring Statistics</u> has been written with you, the student, in mind because it is important for you to be able to read the book and learn from it. The concepts are introduced with real-life examples to give you real, not contrived, insight into the discussions which follow. Begin with the Statistical Insight at the beginning of every new chapter. The Statistical Insights will help you to get "hooked" at the beginning because they are real-life, controversial situations. Relate the ideas to any personal experience and discuss points of interest at the dinner table...yes, I said dinner table. You, as the main person in control of your learning, must discuss the ideas with the same ease as you would an article in the newspaper or a rumor around town because, in fact, most of these statistical insights are taken from newspapers or professional journals. Decide, now, to become interested. Discuss the problems, get comfortable, and enjoy as you work with others and as you see for yourself that you can learn this material and feel good about it. Take the initiative to get started in the book. Take thirty minutes in the first few days of class to read the Statistical Insights at the beginning of each chapter. As you browse through the book get involved with the problems. Take thirty minutes to read a cross section of problems from different chapters of the book. This will give you an overview of the course and will also intrigue you and get you involved. This involvement is essential for learning.

Preview the book

Get comfortable reading it.

Each individual concept is developed in easy-to-read language, yet at the same time the author is thorough in the development of each concept. It is this thorough development that helps you to see all of the puzzle pieces which are fundamental to understanding. Don't look for cook-book approaches to memorize. Don't blindly memorize any formula. Get rid of this approach: "Just tell me what to do and I'll do it." You may get some right answers but with the lack of understanding in that approach, you won't learn, retain, or enjoy anything. Memorization of vocabulary is important when new words are introduced, and memorization of some formulas is sometimes helpful, however, this helpful memorization only comes after the ideas that generate the memorized terms or procedures are understood.

Get interested. Think, as you read the book.

Another refreshing plus of this book is that problems are taken from real-life studies and controversial issues. Allow yourself to become involved with the problems. Be a student of statistics. Question the why's and what if's. Be a statistics private eye. Get inquisitive: be intrigued.

As you go through the chapter, continually review by making a sample test of problems which your instructor has emphasized. Also use the chapter-by-chapter synopsis and sample tests provided in Part B of this book to review the important issues, definitions, then test yourself at the end of each chapter.

4. First-Day Survival Kit

If this is your first day, then great, start here with enthusiasm and determination to begin looking for techniques which make learning easier. Remember you **can** generate enthusiasm and determination if you know what to do to learn. If you are already in the class and looking for ways to improve, compare the first-day ideas to what you are currently doing and make some changes.

> *I'm scared to death. What works for me in other classes doesn't seem to work in math-related classes. Could you just give me some suggestions?*
>
> *~Kristi*

What to do on the first day of class

What you do on the first day of class is really what you do everyday of class. You see, no matter what your track record is, that is, no matter how you've done in previous math-oriented classes, you can start over now, beginning with behaviors which suggest that you are and will be a successful student. Even if you have failed at math-related courses in the past, vow that this time around you will try a different approach and be successful.

Arrive early, sit up front if possible and get ready for class with pencil and paper. Begin meeting those around you and you'll probably find that they are just as apprehensive as you. Relax. Make the instructor feel welcome because even though you may think that you're the only one nervous, think about it, your instructor has to speak in front of a group of 30 strangers and just may be nervous also ... yes may be even more nervous than you are!

Always be on time.

Give yourself time to get ready for class.

Listen and think as you take notes. Don't just write everything down. Listen to what the instructor is saying, think about it as the discussion goes on or as the concept develops, and then (as a second priority) write whatever helps you to recall the concepts later. Re-think these concepts from your notes as soon as possible after class. In taking notes, often you will think and copy the solutions to problems which are worked for you. Other times you will think and write only the key words which suggest the key ideas. You will note any problematic warnings or cautions of typical mistakes that the instructor may point out. **The most common error in note-taking is writing without thinking, with the intention of deciphering the notes later.**

Always follow the thinking as a first priority.

Ask questions. Be vocal. Let the instructor know what you don't understand. Also even if you're only moderately sure of an idea, have the courage to speak up and say 'OK, let me see if I have this right...(and summarize the concept as you see it)." This helps the whole class because the instructor can critique your summary, and the class is encouraged to speak up. If you feel lost and don't know what to ask, make some gesture which suggests to the instructor that the ideas are not clear. Keep asking questions in and out of class even days later after a concept is covered. Keep in mind that you will not always understand concepts right away. Sometimes it takes studying the same idea again and again. Sometimes it takes going on to the next concept before a former one makes sense. Experience this learning-over-time calmly: don't allow yourself to become frustrated and impatient. Returning to an idea and studying it again for more complete understanding is a normal part of the learning process and should be expected. Too many students give up much too soon and decide "I just can't do this. This stuff is too hard for me." Don't let yourself think that. **Never give up on a concept, instead, keep thinking and asking questions.**

Speak up in class.

It helps the instructor to know what you need.

What to do when you leave that first class

Get organized. A good beginning is an organized beginning. Get a notebook with paper and dividers. Put your syllabus and any schedules distributed the first day at the beginning of the book. Include also a sheet of your own on which you will record all grades from major tests, quizzes, homework or any other category which the syllabus indicates will determine your grade. (As you go through the semester keep your grades on this sheet.) The first section is for class notes. Begin with the date and the corresponding section in the book written at the beginning of each day's work. The next section is for homework. Notate clearly the section and page number of your homework. The next section is for returned tests and quizzes which should be reworked routinely. The last section is for supplementary worksheets from class.

Study your notes as soon as possible. Even as you walk out the door, talk about the ideas presented in class. As you walk across campus with another student from class, or as you drive home, mention the ideas from class. Don't think of statistics class as "out of sight—out of mind." Don't shut out the ideas just because you shut your notebook. Keep them in mind. These mental notes which you continue developing as you leave class are part of studying your notes. Sometimes in an informal way after class, a discussion with a classmate will answer a question from class. This is an important part of learning. Look at your notes (mental or written) ASAP after class even if for only a few minutes. Later, when you sit down to study for a longer period of time, study your written notes from class thoroughly. In studying your notes, re-work every problem which your instructor worked and compare your work to that which you copied from class. High-light definitions and theorems and try to write them from memory. Complete any problems from the lecture notes that are incomplete.

Begin your studying by thinking through and re-working your notes.

Read your textbook. Read slowly and carefully. Work every example with a pencil and compare your work to that in the book. When finished, summarize the ideas in the book and in your notes. The two sources—text book and notes—should dove-tail. Memorize all vocabulary. Be able to write it. Also use Part B of this book for a comprehensive list of the main issues and a comprehensive glossary of terms.

Book and notes should compliment each other.

Do your assigned homework. Do homework problems at the end of each section without referring to the examples in the book or to your notes. Referring to the book should become minimal. Continue working until you can work problems without referring to the book. After all, one objective in learning is to retain the material and also on tests you will be asked to work problems without using the text, so the sooner you get used to working totally from your acquired knowledge, the better.

Talk statistics. Find the time to explain what you're learning to another person like, a roommate, a parent, a friend, or a classmate. Get used to talking about the ideas showing that you understand them. Many times as you explain what you're learning, questions arise, which when answered, help you understand the concepts in more depth.

Spend 15 minutes looking ahead to the next section. Each day, no matter how late at night it is, even if you only read the titles and subtitles or a few interesting examples, look ahead to what is coming next. This preview will prepare you for class.

In Summary

- **Think with the instructor in class.**
  (Write down whatever will help you re-think the ideas later.)

- **Study notes and book as soon as possible after class.**

- **Do the assigned homework.**

- **Talk Statistics.**

- **Preview, preparing for the next class.**

5. Your Mind-Set For Learning Statistics

But math-related courses have always been difficult for me! Am I just lacking in ability? My teachers have always said "don't worry, you're an English-history type. But now I have to pass statistics!

~ Debbie

What is a *mind-set* for learning?

Your mind-set is your complete approach to this course. It involves how you see yourself in the big picture of taking statistics. How you see yourself in the big picture involves your attitude, reactions, how you begin that first day, and how you continue through the course. It is primarily tempered by your beliefs—beliefs concerning the subject matter and beliefs about yourself. Become aware of yourself in the three areas—attitude, reactions, and first-day study skills which we have already discussed. Plan to make the study skills given above in "the first day survival kit" habitual throughout the course. As you establish the routine suggested in the "first day survival kit," look at your work ethic, your consistency and your determination as the course continues. Your mind-set also includes your know-how in studying statistics. Your preferred thinking style in this mathematical-statistical course is also part of this. The "way you think" in other courses may not be fruitful in this course without some modification. So, as you acquire new ways of "doing" in this class, begin to see yourself more positively. Your mind-set will become that of a person who feels better and better about statistics. Your "new" mind-set will begin on the very first day with some changes in approach and with different techniques for studying as given in the "first-day survival kit." From that point on, it's a matter of observing your own ways. Examples of realistic and unrealistic expectations, ways of studying, ways of thinking, and beliefs will be given, but you should observe your own "ways of doing" and your own reactions and see if they mesh with the ideas given in this book.

Your objective with the material which follows is to:

- become aware of your own mind-set,

- find techniques for studying and thinking, and

- become aware of the impact of your beliefs on your success.

Mind-sets about math in general

Many people have misconceptions about how to learn math-related subjects. These misconceptions can lead to frustrations because they involve expectations that are not productive. In contrast, consider the following examples of productive expectations and make them part of your own mind-set.

Expect to learn _new_ ideas every class. This seems so obvious, because why else would you be taking the course but to learn something new! But some students remain fairly comfortable as long as the subject matter is familiar, and when it becomes unfamiliar, they become very uncomfortable almost as if they think it's unreasonable to learn new things and be "stretched." Realize that you should expect the material to get harder. It, most probably, will go through up's and down's of easier then harder, then easier, then harder. Expect to study the harder parts again and again as you study new material. Expect your instructor to push you, challenge you, and stretch you—after all, all good coaches do! Realize that you will be taught ideas that you have not seen before in school. Don't get nervous or scared at the introduction of new vocabulary (Statistics is full of big words!). The nice part about the vocabulary is that it generally means what you think it should mean from everyday usage.

> **Don't balk at new ideas; Instead,**
>
> - **Take a deep breath,**
>
> - **Listen and think in class,**
>
> - **Read the book at home,**
>
> - **Discuss the ideas outside of class.**

Expect that learning will not be instantaneous. One of the most debilitating expectations that students have is that they will understand the ideas completely when they are presented in class. This expectation is unrealistic because learning statistics requires patient and thoughtful studying, working and reviewing problems sometimes again and again over a period of time. If you do pick up on an idea quickly without much effort, then you must still practice it. Challenge yourself to work problems in the book noting any modifications. If a topic is not clear in class, then you must study it in your notes and in the book and return to class with questions having tried your best to understand and work problems. You must also expect that, at times, you will spend time with a concept without a complete understanding of it. If you know that you are impatient in learning, try to replace the expectation of immediate understanding with this expectation: *The book and the class presentation will only introduce ideas. These ideas, must be read and re-worked in notes and the book. Practicing the thinking by working several different examples and homework problems and connecting the ideas to the previous work is imperative.*

This expectation is important because if you expect instant gratification and don't get it, you may begin thinking that you can't do statistics (which is a belief and part of your mind-set). On the other hand, if your expectation is to think and re-think the concepts before you understand, then you will stay psychologically strong and healthy.

If it's been a long time since you've been in this type of course, just have patience with yourself. Study. Let yourself make mistakes. Allow yourself to learn the material. If you're a perfectionist, change that expectation. You're not expected to be perfect, you learn from making mistakes. Thinking that you have to make the highest grades may really stress you and be harmful to your learning. Just be determined to learn from your mistakes and realize that you learn by speaking up in class and by revealing your stumbling as well as your insight. Nobody expects you to know the course before you take it. Learn from your mistakes, be easy on yourself and progress.

Affirm for yourself:

I will study daily and continue thinking about ideas which are not as yet clear.

Your learning is like a snowball rolling downhill—it constantly gathers snow and becomes bigger and bigger. Sometimes the hill flattens out and the snowball seems to slow down. Sometimes the snow doesn't seem to stick as well, and sometimes there's not much snow. But at other times the growth and the speed is awesome. But, no matter what, it will continue to grow as long as it keeps on rolling.

Expect varying backgrounds, strengths, and weaknesses from the students in the class. What if you feel terribly unprepared for the course, what if you feel like you're struggling more than others.

> *I feel so stupid because in other areas I am confident, but here I feel embarrassed by the multitude of questions that I have to ask.*
> ~Jamie

Feelings like these are often intensified by unrealistic expectations you hold for yourself. Let's face it, everyone in the class cannot be at the same level. You will have younger, older, quicker, slower, intuitives, logicals, and there will be a host of other differences which actually make people individuals and should make the class more interesting. Be proud of your strengths, bring them to the class, and don't be embarrassed at your own feelings of weakness. The best thing to do is drop this embarrassment and just ask questions, work together, and learn from one another. You will have strengths which will complement others' strengths. It's like placing your cards on the table showing the hand you've been delt and then working with others to get the best hand you can. You can look at individual differences as a normal part of life, get comfortable with these differences, and work with others. As you learn and proceed, these differences will be lessened because you'll learn from each other.

What if you do feel like a "fish out of water?" Think about why you feel that way, and do something about it. Change your self-talk to focus on your strengths. A willingness to: read the book, think about the concepts, work hard, meticulously keep up, and organize a group of students to work together will help you become more comfortable. Study meticulously by yourself and in groups. As you work together in your group, practice reading the directions and the problems carefully to determine the method to use on problems. Understand how and why the concept is applied. Then, and only then, work the problems. Don't just go through the steps of working problems. Groups can be very effective if each person takes the leadership role at different times, discussing what you do and why you do it. Have each others' phone numbers because quick calls can be very helpful when you're stuck. If you still feel uncomfortable, continue changing your self-talk to positive. Remember, if you're studying hard and at the same time learning how to think a little differently, it will take time. It's like trying to move a rusty wheel. It takes some jarring, a lot of oil, and some working through the rough spots, but soon, with use, it smoothes out. Tell your instructor, as well as the other students in the class, how you feel. Let them respond. Sometimes your instructor is at a better perspective than you to evaluate your progress, and your instructor's evaluation may surprise you. You may feel like you're struggling and just hanging on by a thread, and your instructor may say, "trust me, you are working hard and it will come together for you as you continue." She may also take the opportunity to give you some study suggestions to fine-tune what you're already doing. If you and your classmates share your feelings about the class you will find that you're not the only one feeling the way you do.

Expect that the algebra skills that you need to be successful at statistics are minimal. What you do need are some very basic algebra techniques and the maturity that comes from being successful in an algebra class. If you've never had algebra before or if this feeling of basic immaturity is there then perhaps you should take a beginning algebra class. On the other hand, if you can get involved with the ideas of statistics and not let the unfamiliar notation and vocabulary bother you, but instead, make sense of it, then you should proceed with the course, and review the algebra techniques as needed.

Expect to study right away, every day. Whether specific homework is assigned or not, your instructor expects you to study. Don't expect your instructor to tell you every little thing to do. Study—spend time thinking and questioning the concepts, and you will learn. Procrastination will absolutely undermine your efforts at success. The more time you put in studying right away—as soon as possible after class—the better you will understand and retain the material because as the instructor continues developing the ideas perhaps over several days, each new idea will dovetail with previous ideas. You must stay current on the ideas, question them, relate them to other ideas, and discuss them with others in the class.

Daily-study check list

Are you doing the following daily? Respond with "yes" or "no." Do you:

1.   Preview the up-coming material in the book?   _____

2.   Go to class daily—ask questions, take notes?   _____

3.   Rethink the notes and hear the instructor's voice in your head as you think?   _____

4.   Read the book—work examples with a pencil?   _____

5.   Try to make the notes and the ideas in the book to dove tale?   _____

6.   Talk statistics?   _____

7.   Do homework problems without referring to notes or book?   _____

You may be thinking, "I don't have time to study this way!" Well, this is the most time-efficient way to study. You may think that reading the book and studying your notes is too time consuming, but many times the primary learning occurs with that exercise. Then, after studying your notes and the book, the learning process continues as you work the assigned problems.

Take the time to study thoroughly.

You'll save time in the long run.

Schedule your time wisely. Set aside time in your daily schedule to study. You should have a minimum of two hours out of class for every hour in class to study thoroughly, but that two hours does not have to be in one block of time. For instance, studying ASAP after class may be on the bus on the way home, in your car as you commute, between classes, as you workout in the gym, as you eat lunch, as you're waiting on someone. Adapt this to yourself. Find ten-minute time intervals to think about the ideas from class. Put ideas on readily accessible index cards. Discuss the ideas with classmates, as you walk from class, before class begins, on the phone, over lunch. Take your book with you wherever you go and read it. The point is, you can read and think about the ideas at times that you would have never thought to do so, and it is gratifying to use your time wisely. Also schedule a block of time specifically for studying statistics. At first, you may have to remind yourself to study, but gradually studying will become automatic as you begin to see the benefits.

Try the above suggestions for a few weeks or until the first test. If your test score is low, go to your instructor with the following: your test reworked, your class notes, and your homework. Study any concepts that your instructor re-teaches you. Also, discuss your study techniques by letting your instructor listen to your description of how you study and offer you suggestions.

Check out your attitude toward study skills

Circle from 1 to 5 where:

1= not at all
3 = moderately so
5 = I understand fully

1.	I understand the importance of wanting to learn statistics.	1	3	5
2.	I understand that it is normal NOT to understand a concept instantly.	1	3	5
3.	I understand that it is normal to be puzzled by a problem, take a break, return to the problem, and then understand it.	1	3	5
4.	I understand that I may not understand a concept even after studying it for several days.	1	3	5
5.	I understand the importance of beginning to study concepts on the same day that they are covered in class.	1	3	5
6.	I understand that procrastination will undermine my efforts.	1	3	5
7.	I understand the importance of studying daily, thoroughly, and not procrastinating.	1	3	5
8.	I understand the importance of asking questions and do so regularly.	1	3	5

Total _____

Any questions that you marked (1) or (3) should be discussed with your instructor. As you go through the course, answer these same questions again and again. Monitor your progress. Your total of all the numbers circled should go up as you try the suggestions in this book, as you study, and as you learn statistics.

6. Your Mind-Set For "Thinking Statistics"

What is your approach to thinking in math-related courses and in particular in this course? Have you ever thought about it? Is your thinking logical, sequential, and stepwise or, on the other hand, intuitive, impulsive, and creative? This is an important question, and if you've never thought about it, it is important that you analyze your thinking in this course, other courses, your former math courses, and even non-school activities, because you may be spending hours with intuitive guessing and mimicking without any logical thinking.

Think about your own approach to thinking

What difference does it make to be logical and sequential—a left-brain-dominant thinker vs. intuitive, creative, and impulsive—a right-brained-dominant thinker?

Students who prefer to think more logically and sequentially (LB) will, quite naturally, read directions, write the problem, work the problem showing steps, and stay on track to completion. The students who are naturally more intuitive (RB) sometimes have trouble keeping on track with a problem. The logic may be vague and the steps taken may be a result of guessing or mimicking what was done in class. They may spend a considerable amount of time "doing work" without really thinking and learning the concept. Their lack of emphasis on starting at the beginning and progressing through the work also makes it difficult to routinely study the notes from class, read the book, and do the homework problems.

Characteristics of left-brained and right-brained thinkers

Left		Right
logical	vs	intuitive
sequential	vs	impulsive
set-by-step	vs	focus on the whole
meticulous, painstaking	vs	imprecise yet quick

The desired way of thinking is actually whole-brain thinking—a mix of the logical/sequential (LB) with the intuitive/creative (RB). Students, in a right-brain way, can get involved with the Statistical Insights at the beginning of each chapter and other interesting problems in the book but also in a left-brain way take the time to understand the development of the concepts, memorize the vocabulary, and work many problems rethinking the logic in every step. The focus, here, is not to question which side of the brain is responsible for what. The focus is to realize that there are two distinct ways to approach thinking— logically vs. intuitively. Many people have a preferred approach while others more naturally use a combination. If a student uses intuition excluding logic to an extreme, he will have trouble. Intuitive thinkers have particular problems when they cannot see the solution to the problems in their minds, because without step-by-step logic they generally have no place to begin working the problems. On tests this is particularly frustrating because they literally cannot begin the problem except by guessing. It is quite possible and desirable to become more balanced, with logic being the most important element.

So, what is your natural, preferred approach to thinking? Think of something that you do well. Can you find a step-by-step sequential element as well as a creative element? If so, you have a whole-brain approach. List several things that you do very well and think about your approach to thinking

_____, _____, _____, _____, _____

such as cooking, sewing, carpentry, musical performance, dance, sports, art, mechanics, writing, etc. There is a planning stage, an outline, a learn-the-notes stage, perspective, a learning the function of the parts, etc. which is the left-brain component, and there is also a creative stage or a picture-how-the-whole-thing-goes-together stage which is the right-brain component. Usually the planning and nit-picky studying or practicing goes first before the creative genius, but it doesn't have to. The important thing to note is that, you plan, create, picture and get a result. Perhaps, the degree to which you meticulously plan, is dependent upon how perfect your final result has to be. Reflect on the following: When you cook, do you follow the direction in the cookbook to the letter(LB), or do you just throw things together(RB)? When you write, do you outline (LB) or do you begin creatively (RB) and at some later point check your thought development. If carpentry is your hobby, do you study the plan (LB), or do you see in your mind or in a picture what you want to build and just start (RB). In music do you read notes(LB) or do you play by ear(RB)? If you play a sport, there is the drill and learning part (LB) and an intuitive and impulsive part (RB) when you finally "play" the sport comfortably.

> **In the things that you do well, the logical (LB) and the intuitive, creative, picturing (RB) work together to bring about a whole-brain process which gets better as you continue to work at it.**

Statistics has wonderful left and right-brain components. The right-brain sees the whole picture and relates to the meaning in the situations in the problems. The left brain is concerned with the logical development of the concepts involved in solving the problems. Both are important and compliment each other. Your instructor will probably discuss statistics problems and their relevance (RB) and will also show the details in solving problems on the board (LB). Acquire LB and RB techniques by:

listening to your instructor (who will model LB/RB thinking), reading the book, working by yourself (imitating your instructors thinking), and working with others (thinking through problems together).

If the above discussion makes sense, then it is time to consider your thinking style in this statistics course. Reflect on what your thinking style has been in past courses. Also begin observing your thinking style in activities which you excel in school or outside of school. Knowing your strengths, and knowing your thinking style in your favorite activities as well as in this class is very important to your becoming aware of how you need to think in your studying to achieve success.

Suppose you are extremely right brained and just can't get started in meticulously going through the problems. Well, don't avoid the problem, but instead, hit it head on. Organize your notes, homework, and tests. Work cooperatively with others to develop your LB skills and, at times, take the initiative to lead the discussion and explain concepts. If you have trouble seeing the relevance of the development of theorems or any of the concepts which the instructor shows on the board, discuss this thinking with another student after class. Unfortunately, many students begin their homework by going straight to the problems, which may yield incomplete understanding of the concepts. Don't begin your studying with homework problems without first studying the concepts, because you may be skipping over the most important part—the development and understanding of the concept.

No guessing allowed! Practice logical thought development with others. Every step has a reason. Don't mimic steps without thinking "why." Only make moves because it makes logical sense to do so. Also, question every little detail. Also realize that if you struggle with this process, it is not due to an inability but rather an untapped thinking style. If you begin to understand your thinking preference and learn to think logically and practice this thinking you will get better at it. It will compliment your more intuitive strength of picturing the problems and ideas and you may begin to really enjoy statistics.

Also don't let the notation of statistics frustrate you. Statistical notation is tight and precise and that is much more appealing to the left-brain. Have patience with it, study it, write it, learn to use it, and you will learn to appreciate it.

Your mind-set of self—your beliefs

The acquisition of negative beliefs may or may not be related directly to your preferred way of thinking. Many logical (as well as intuitive) students have worked hard in a math-related course, not made the grade, and finished the course discouraged and feeling that they just can't do the course work. Because of this, their mind-set is clouded with feelings of self-doubt. They have felt that it is an ability-related problem and once they feel very strongly that they can't do something (a belief), no matter what it is, they have a hard time succeeding at it. If you feel like this then this next material will help you understand what to do about it.

> *How can past experiences and beliefs possibly affect me now? I am determined to succeed.*
>
> *~Jan*

Beliefs are an interesting component of your mind-set in all aspects of life. What you believe you are, you'll be. What you have affirmed in the past (like "I hate this," or "I'll never do this again") will affect you unless you make an effort at changing the affirmation. If you "know" you can't do something, you won't be able to, or if you do succeed, it will be with a struggle. Students come to statistics courses with debilitating beliefs as a result of their past experiences in math courses. These negative beliefs more than anything guide present performance and sometimes lead to failure. In a similar manner, positive experiences lead to positive beliefs and success. Fortunately, however, beliefs are learned and can be unlearned. Awareness of your beliefs and how they affect you is the first step toward acquiring positive beliefs and success.

How did you develop negative beliefs? Acquiring a negative belief can happen very gradually and subtly over time or in an instant with one bad experience with a parent or teacher. In math-related fields, more so than others, you hear of students deciding that they are weak and harboring that belief from then on. Some students don't study enough, some don't know how to study. Some have bad teachers. Most are not aware of the left and right brain components in learning. The early learning of math can be so dry and perhaps memorized that many are disenchanted with the tedious nature of elementary-school math. Some students just stop working, and then find themselves behind and relying on others for homework and grades. Society even dictates myths like: math, science and engineering are male-dominated fields. Some parents, without really realizing it, have passed along beliefs like: "I couldn't do math, so you can't either." Students can relate traumatic experiences from the classroom where they have felt humiliated by certain teachers and decided, "I hate this," I can't do this," or "I'll never do this again." Another source is from home when a parent has punished a child while the child was trying to learn math. Well, now it is important to realize that those affirmations made years ago may be having an effect on your attempt at learning any math-related field including statistics. You may find yourself in a negative syndrome which has gotten stronger over the years.

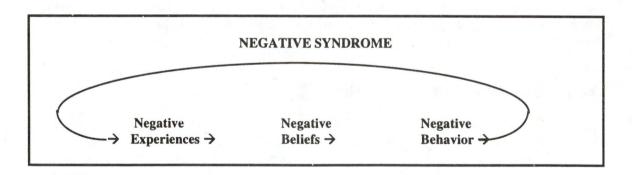

NEGATIVE SYNDROME

Negative
→ Experiences →

Negative
Beliefs →

Negative
Behavior →

Now, you have to break out of the negative syndrome. But how? How can you acquire positive behavior if your past has been filled with negative experiences yielding negative beliefs resulting in negative behavior?

How do you take on positive beliefs? Change something. To take on positive beliefs and break the syndrome, somehow, have a positive experience. Take on the suggestions in this book. That alone will be positive behavior. Begin the first day. After you are aware of how the syndrome works and where you are in the big picture, begin changing. Begin positive self-talk and begin believing in yourself.

Change your attitude and make these ideas work for you. Monitor your reactions to the up's and down's of class. Change study skills in the realization that perhaps in the past you had no motivation to work, and really didn't know how to study. Take on logical and sequential thinking, get involved with the book, and develop a more whole-brain approach. Also bear in mind that nothing will work for you if you don't put forth 100% effort in learning the material.

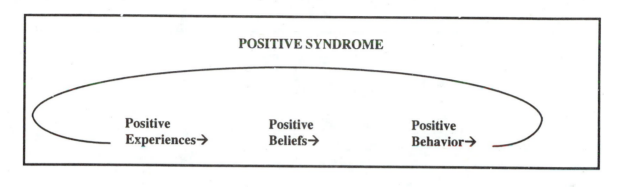

POSITIVE SYNDROME

Positive
Experiences→

Positive
Beliefs→

Positive
Behavior→

At this point write at least five suggestions you plan to take from this book which will help you break into a positive syndrome.

_____

_____

_____

_____

_____

What else should you do? Relate your beliefs to your learning preference. Think about your preferred thinking style and how it has affected learning and how it may have been related to your apparent inability and thus negative beliefs. Realize why math-related course work has been a problem for you within the context of your beliefs. Reflect on your past and question your beliefs. Then as you begin changing, begin telling yourself, "I'm going to be successful this time. I can do this, because now I know how." Squelch negative self-talk and replace it with a positive counterpart. And, I can't emphasize it enough, **work hard**, harder than you ever have before. Remember, working hard is a characteristic of strong students. Begin telling yourself "I am becoming a strong student." Your mind-set will come together for you little-by-little as you work to learn statistics.

Start noticing your self-talk. Use this self-talk as a clue for finding your beliefs. Also, through conversations with others who know you, uncover beliefs. Then, question the validity of beliefs. Begin taking on a belief "I'm just as smart as the next person." Actually, you may even be harboring a negative belief like: "If I have to work that hard then I must really be stupid" or a belief like "Only slow people ask questions." Change these beliefs! Most teachers will tell you that their best students study long hours. They'll also tell you that learning occurs through making mistakes and good students ask an abundance of questions. You see, your beliefs may be totally unreasonable and may be keeping you from a behavior which will lead to success. If you and other students look for beliefs in each other and question them, you can help each other shed negative beliefs and replace them with beliefs that are productive.

Find negative beliefs.

Question the validity.

Replace negative beliefs with positive beliefs.

Summary of your mind-set for success

Your mind-set involves four inter-related components:

- attitude

- study skills

- approach to thinking

- beliefs.

Find yourself in the above components and develop a positive and effective mind-set.

7.   An Experimental Mind-Set—How You See Yourself

One imagery that you might try, particularly if you are terribly fearful of this course is to temper your mind-set by mentally taking yourself out of the classroom. Don't see this course as a *school* course, but see it as your employment. Your instructor is your employer and you have a job to do. You, quite naturally, ask questions, and do whatever you're asked to do. Your job is important. As an individual who has perhaps had different jobs in the past, you are trainable for this job. You will advance only if you apply yourself, care about how things work, and take on the suggested techniques as part of your everyday work. You will be rewarded for questions regarding the inner-working of the job because when questions are raised, answers help make the product better. Success breeds success, and you will discuss interesting finds and questions with other employees. Financially, you have stock in the company and the better you do, the more valuable your stock becomes. Enthusiasm will grow as every employee seems to have a vested interest in the company's success.

Problems, questions, ideas, difficulties, challenges, and set backs will be seen as a normal part of everyday work. Also, every challenge presented by the employer or the factory manual (the book) will be considered thoughtfully until thorough understanding and connections are made with the already established processes. Your boss will not watch your every move. He trusts you to do the job understanding the techniques which he is showing you. You will be rewarded as a result of studying, working, and thinking about your work no matter what is asked of you.

You will go to work everyday because people are depending on you. Production will slow down if any one person is missing. You will go to your boss with ideas, concerns, questions and you'll learn how to do your job better from every correction or adjustment which your employer makes.

Notice that there will always be employees who are quicker at understanding or doing, but it doesn't matter because each person has a place and is important. Your work will become better as the whole team works together. "No man is an island" in this company. Every person brings a different strength or talent to the job. You are no exception. Do your part! Tell yourself "I can be successful at any job that I am hired to do."

In summary, see this course as your employment and do your job. This isn't school, this is your employment. Understand what the employer is asking you to do and don't stop asking, thinking, and commenting until the work you're doing makes sense. Ability is not a factor. What matters is your hard work and determination to do your job well.

8.   Studying For And Taking Tests

If you've had trouble taking tests in the past, now it's time to hit the problem head on within your new mind-set.

> *I am not good at taking tests. I study, but I just can't do the tests. I think there should be a different way of giving grades.*
>
> > *~Amy*

First of all, if your studying techniques are changing and your approach to thinking is becoming more logical and sequential then your test-taking should become more logical and sequential. Giving and receiving grades is not the main reason for tests. Tests are part of the learning process and not ends in themselves. You will take tests with the objective of showing yourself and your instructor what you know and don't know. Don't be too quick to call your mistakes "careless." The primary purpose of taking a test is to learn from your mistakes.

> *If test-taking is a normal and expected part of the course then what makes me so sick?*
>
> > *~ Amy*

Some of the nervousness is expected and good. It's the same nervousness you feel before a performance or a game. Debilitating nervousness is what you want to eliminate. The primary cause of this nervousness is not understanding the concepts and consequently working intuitively, guessing. Not being able to take tests successfully is generally a signal that you're not learning—really thinking and learning—the material. That is not to say that you're not putting time and effort into the course. It may be that you are putting a lot of time in but not learning the concepts.

The most common reason that a person experiences test anxiety or does poorly on tests is that the concepts are not really learned, understood, and then practiced.

If you are taking on a new approach to thinking logically and studying hard to really learn the material, then your test-taking skills should improve.

What specifically should I do for a test?

The best preparation for a test is the daily studying described earlier. The following suggestions refer to the preparation during the few nights before a test. If your studying has been sporadic then the cramming you do at this time is really not very beneficial. If your studying has been regular, then what you do at this time is review, practice and get psychologically ready.

Make yourself a test. Make this test daily, beginning on the day following your last test. After each class lecture, add to your test., including problems from lecture emphasized by your instructor and from the book. Take your test every other day even though in the beginning you will only have a few problems on the test. In this way you will review concepts every other day as you go on to new material. When the test comes you will have reviewed regularly. Trade these self-tests with others in the class. If your instructor chooses to review, then add to your test with some exercises from the review.

Review. Re-study the development of the concepts and examples in the book and in your class notes. This may seem time consuming yet it is the most worthwhile part of studying for a test because you will see the chapter as a whole. Summarize the major points of the chapter. When you feel that you know the chapter, turn to Part B of this book and use the comprehensive list of Important Issues and the Glossary of terms to see if you have reviewed thoroughly. Be able to write, from memory, important definitions and formulas even if you are not expected to have them memorized. Answer the Review Questions in the textbook and re-study the concepts that correspond to the questions missed. Also work more exercises at the end of the appropriate sections. Then when you feel that your studying is complete, work the practice tests provided in Part B of this book.

Psychological preparation

Imaging, Changing beliefs, and Positive self-statements. Imagining involves picturing yourself doing things that a successful student does or things that you do in classes in which you are confident. This includes listening in class, asking questions and participating, studying, doing homework, discussing concepts with others, or taking tests in class. Imagine yourself in another class in which your behavior or performance is successful. Mentally note your frame of mind. Then instantly, imagine yourself in statistics class with the same frame of mind. "See" yourself participating in class, or studying with others, or taking the test with the same ability to perform as in the other class. In sports, coaches use imaging techniques asking their players to "see" themselves swinging the bat level in baseball or using the proper form in shooting free throws in basketball etc. What you imagine in your mind, you can be.

Another example of imagery is to picture learning statistics as learning a card game, a board game, or a sport. When you're learning a game, you are comfortable asking "How do you know ...?" or "But what if...?" You are interested, have fun with others, and have the satisfaction of thinking and playing. See this course the same way. You can change your perception because you are in control.

Another related technique involves telling yourself, as you go to sleep, that you have done everything suggested (enumerate what you have done to study on a daily basis) and that you will do well. Form the belief that "I can do well and I will do well," then go to sleep confident that you have studied properly and thoroughly. Do not cram the night before a test or the next morning. Tell yourself positive statements ("I can ... and I will...") and relax with deep breathes whenever you begin to doubt yourself.

Another suggestion to use in preparation for a test is to develop and use positive self-statements any time you begin to get nervous about the test. This may be weeks before the test as well as right before or during the test. Weeks before the test you may have knots in your stomach or experience an inability to sleep or just feel nervous. At times like these, take deep relaxing breaths. Breath away tension. Let yourself relax and tell yourself positive statements. Choose from the following list any statements which seem appropriate for you and say them to yourself at the first sign of tension.

In the days prior to the test, say to yourself:

- I am studying properly, so just relax.

- I've done everything suggested, I'm ready.

- If I make some mistakes that's normal. I'll learn from them.

- Making an A is not important. What is important is to do my best and learn from mistakes.

- If I do poorly, I'll discuss the test with my instructor and use the suggestions from the conference.

Immediately before and during the test, say to yourself:

- A test is a game. My primary objective is to show myself how much I know.

- I have studied the strategies for each problem. Relax, read directions, and base my answer in an underlying concept. Use this thinking on each problem: "What is the first step and why?"

- As the test is handed to me, I'll take a deep breath and concentrate on my test. I will not expect perfect quiet even though I would prefer it.

- If another student leaves early it won't bother me. Leaving early may mean that the person knows very little.

- If a problem doesn't work out nicely, I'll develop it as completely as possible, then go on.

- When I don't know what to do on a problem, I'll develop it as completely as possible, then go on.

- When I don't know what to do on a problem, I won't give in to the temptation of just "doing something." Instead, I'll skip it and come back to it. No guessing!

- If I feel myself getting flushed with anxiety, I will take a deep breath, say positive statements, and look for a problem that I know I can work.

- If the format of the test is not what I expected, I will not get upset. I will take a deep breath and continue working. After all, if I know the material, the format shouldn't matter.

- If I get overwhelmed with anxiety or anger, I'll say to myself "If I continue like this, I'll surely fail. So relax, begin somewhere, anything is better than what I have now."

Make up your own statements.

- _____

- _____

- _____

- _____

Complete your studying by telling yourself, "I'm ready."

The mistake some students make in studying for a test is in not reviewing the development of the concepts in an organized fashion and in not studying all of the concepts in the chapter. Instead, they merely work problems. They may begin with the Review Questions and if unable to work them they look up the answers and try to decide how to get the answer given. This is not the way to learn or to review for a test because the foundation is weak. You must re-study the entire chapter first, then work sample problems (Review Questions), then test your understanding of the concepts with the Practice Tests provided in this guide.

Watch your self-talk during the test. Change your self-talk if it is negative. Be as kind to yourself as you would to others in the class. You would never let fellow class-mates talk about how stupid they are, or say "I'm going to fail" after you've studied together for hours. Don't allow negative self-talk in yourself or others.

9.   Troubleshooting

Suppose you just can't make yourself preview the book. You can't seem to make yourself attend class regularly or study the notes or read the book or think about the statistics,

**THEN IT'S TIME TO REFLECT MORE
DEEPLY ABOUT THE PROBLEM.**

So where can you start to look for answers to understanding your behavior? First, question your goals. Do you have a major? Do you know the type of career you'd like to have? The more direction you have, the more motivated you'll be. If your goals are tentatively set, and you don't feel that a lack of direction is your problem, then look at your beliefs in more depth. Do you have some hidden agenda for your counter-productive behavior?

Ask yourself: "Would I act this way on a job? Would I just not show up for work? Would I refuse to care about the work I do? Would I keep this job acting this way?" Just as you control these factors in a job setting, you can learn to control your actions and reactions in class.

Question your goals. Where do you see yourself one year from now?...three years, ...five years? Do you have a career in mind? Is your career choice something you really want to do or is it something your parents want? Would you rather be on a job which is more physically demanding than mentally demanding? Many schools offer career counseling. Also perhaps you should look for an interim job doing physically demanding work. It may make you appreciate the less physically demanding life of studying. What does this course have to do with your goals? Find any connection you can between this course and your goals.

_____

_____

Suppose your goals are established, then keep searching for your beliefs. Take on the suggested behaviors. Re-read "Your mind-set of self—your beliefs." Tell your instructor, a counselor, or a friend how you feel. You can change your beliefs that were set in your past and start over with a clean slate.

10.   Sum It Up

See how well you're doing. How well have you incorporated the suggestions given in this manual into your mind-set? Answer the following questions now and often throughout the course. The "no" responses indicate a need for "fine-tuning" of your mind-set.

Answer these questions:

1.   Have you gotten interested in the concepts in the course?

2.   Have you gotten interested in the problems?

3.   Have you spoken to your instructor?

4.   Do you re-think your notes ASAP after class?

5.   Are you making practice tests as you study each chapter? Do you use the tests in this book?

6.   Do you arrive to class early, with paper and pencil ready?

7.   As you're waiting for class, do you review what you did last class, and preview today's class?

8.   Do you spend 15 minutes previewing the new material the night before class?

9.   Do you work every example in the book on paper?

10.   Do you read the explanations in the book which precede the examples in the book?

11.   Do you know at least one other person in the class by name and phone number?

12.   Do you discuss statistics outside of class?

13.   Have you mentioned at least one interesting problem to a friend other than your classmates?

14.   When you read your newspaper, do you look for examples of statistics?

15.   Do you speak up in class?

The above 15 questions are a reflection of your attitude and study skills. If you answered "no" to any of the above questions, consider a behavior change. If you will not allow yourself to do the above, then begin asking yourself, "why can't I take on simple behavior changes like the ones suggested?" Then look for beliefs which inhibit change.

Continue answering these questions:

16.   Does every step of every problem you work have a logical reason?

17. Do you read the directions on problems and know what is being asked?

18. Do you know when a problem is finished?

19. Do you follow the logic in the problems worked in class?

20. Can you work all of the problems from class explaining them to someone who is unfamiliar with the material.

21. Do you have your notes, homework and tests arranged in order in a notebook?

22. Do you have a schedule, and meticulously stick to studying statistics daily?

Questions 16 through 22 deal primarily with acquiring a left-brain mind-set. If you answered "no" to any of these, try to take on the order which will help you succeed.

23. Is your self-talk positive?

24. Do you comfortably ask questions in class?

25. Do you feel confident that any mistake you make is part of the learning experience?

26. Have you given up the dread concerning statistics that you once may have had?

27. Have you been able to shed negative behaviors and replace them with positive behaviors?

28. Have you tried imagery?

29. Have you dispelled any societal myths that may have applied to you?

Questions 23 through 29 refer to your mind-set—your beliefs. Actually, any of the above questions 1 through 29 could be related to a negative belief. Sometimes it's hard to separate the four-- attitude, study skills, approach to thinking, and beliefs, because your beliefs are the basis of your behavior.

From the 29 questions above, how many "yes" responses do you have? _____ Keep working on the ones you answered "no." Improvement should be gradual if you are beginning to break out of a negative syndrome. Expect continuing improvement and keep learning from your mistakes. Read "Developing A Mindset For Success" several times as the course progresses. Look for suggestions to fine-tune your approach. Study hard and always stay with the course until the end no matter what, because if you keep working hard, sometimes the concepts begin to gel more and more toward the end of the course.

As a last word, I would suggest that you relax, have confidence in your own ability, and get involved in the learning of statistics. In this way you can enjoy your statistics experience.

The Statistics Tutor

Part B

Chapter Summaries with Practice Tests

1

Collecting And Understanding Data

The topics of statistics considered in this book are collecting organizing interpreting and presenting numerical information. *Descriptive statistics* is the means by which the collected data is organized and summarized. *Inferential statistics* is drawing conclusions about a population from the data in a sample.

Important issues—Chapter 1

- You should understand the basic concepts of a population and a sample. In a given situation you will be asked to identify the population and describe the sample.

- Often we distinguish between the *target population* and the *sampled population*. Recall that they may differ, but it is hoped that the main characteristics of interest are similar in the two.

- A *census* is sometimes used when collecting data. You should know the properties of a census and why a sample is usually more practical.

- In a specific application you will be asked to identify the *variables* of interest and understand how they are used in the statistical study.

- The main objective of *data analysis* is to organize and summarize the information in a data set to make it more comprehensible. As you progress through the book you will learn more and more about the practice of data analysis.

- The *variation* exhibited by a variable is of primary interest to the statistician. Furthermore, the *distribution* of the variable illustrates the pattern of variation. For small data sets you should be able to exhibit the distribution with a *dotplot*. In later chapters you will be introduced to other descriptive tools that are used to investigate the pattern of variation exhibited by the distribution.

- The sources of data are many. Often statisticians are interested in analyzing the *pre-existing data* that have been collected by others. For example, information from the decennial census conducted by the Census Bureau is continually analyzed for trends and patterns. Statistician also collect their own data for analysis. In so doing they should

  1. Determine the objectives of the study
  2. Choose variables to be measured

3. Identify an appropriate design for producing the data
4. Collect the data

- *Surveys* and *experiments* are two general areas of statistics that are used to collect data. A survey is studying the opinions of the subjects in an existing population. An experiment, on the other hand, is concerned with a population that has been altered for study by the investigator.

- Samples may or may not be *representative* of the population. *Bias* is a systematic tendency of the sample to misrepresent the population. Statisticians are continually concerned with developing sampling procedures that reduce bias. The *simple random sample* is of primary importance in this effort.

- When critically appraising data we should be concerned with

  1. the source. Is there reason for the source to report biased results?
  2. the collection procedures. Is the sample representative of the population?
  3. the usefulness of the data and whether it is presented properly.

- Things to consider when studying a survey are

  1. the population
  2. the method of contact
  3. the response rate
  4. the questions
  5. the timing
  6. the sample design.

- The different types of sampling techniques are

  1. convenience sample
  2. simple random sample
  3. systematic sample
  4. stratified sample
  5. cluster sample

- An *experiment* attempts to establish a cause-and-effect relationship between two variables. The *treatment* is controlled or manipulated by the experimenter. The *response variable* is the variable whose change, if any, is caused by the treatment. A *control group* is the group that does not get the treatment. An *extraneous variable* is a variable, outside of the experiment, whose effect might be *confounded* with the treatment.

- The *comparative experiment* with *random allocation* to groups is used to avoid confounding. An *observational study* is the study of the effect that a treatment has on the subjects when random assignment is not possible.

30

Glossary

An **experimental unit** (or **subject**) is the smallest entity that is of interest in a statistical study.

A **variable** is any characteristic that can be measured on each experimental unit in a statistical study.

An **observation** is a value that the variable assumes for a single unit.

The collection of observations assumed by the variables in the study is called a **data set**.

The **population** is the collection of all objects, or items that are of interest in a statistical study. The individual objects in the population are the experimental units or subjects.

A **sample** is a finite portion (subset) of the population that is used to study the characteristics of concern in the population.

The **target population** is the population under study. The **sampled population** is the population from which the sample is obtained.

A **census** is a sample consisting of the entire population.

The **distribution** of a variable specifies the distinct values that it assumes and how often these values occur. It illustrates the pattern of the variation of the data.

Bias is a systematic tendency of the sample to misrepresent the population.

A **simple random sample** (SRS) of size n consists of n elements chosen from the population in such a way that all samples of that size have the same chance of being selected.

A **bar graph** is a picture consisting of horizontal and vertical axes with rectangles (or rectangular objects) representing the frequency (or amount) of the categories of a variable. The categories of the variable are listed along one axis and the frequency along the other.

A **confounding variable** is one whose effect on the response variable is mixed up with the effect of the treatment.

An **observational study** (also called a quasi-experiment is an experiment in which one observes how a treatment has already affected the subjects.

Practice Test 1

I. True/False

1. The Hite report would have been more accurate in its predictions if 200,000 questionnaires had been sent out instead of 100,000.

2. The sampled population is always a subset of the target population.

3. Usually cluster sampling is more cost effective than simple random sampling.

4. The comparative experiment avoids confounding.

5. The undecided vote should always be split evenly among the candidates.

6. Two variables are said to be independent when they cannot be distinguished from each other.

7. A response variable is a variable that is measured on each experimental unit to determine if its value is affected by the treatment.

II. Multiple Choice

8. A sociologist wishes to select a sample of 100 from a list of 1000 welfare households. She chooses one household at random from the first 10 on the list. It is number 7. She then takes as her sample the households numbered 7, 17, 27, 37, ---, 987, 997. This is a
(a) convenience sample.
(b) simple random sample.
(c) systematic random sample.
(d) regular sample.

9. A psychologist wanted to study the effect of home environment on IQ scores. She selected a sample of students, gave each an IQ test, and also asked the size of their family, their father's occupation, whether or not their mother was employed, and many other questions. In this study, the following are response variables
(a) father's occupation
(b) family size
(c) IQ score
(d) both a and b above.

10. The response rate of a sample survey is lowest when the survey is conducted
(a) by mail.
(b) by telephone.
(c) by personal interview.
(d) by a government agency.

III. Fill in the blank

11. A _____ is a list of units from which the sample is chosen.

12. _____ is a systematic tendency of a sample to misrepresent the population.

13. _____ is used to produce equivalent groups prior to the experiment.

14. To sample the registered voters in a state and make sure that representatives from each county are included, what sampling technique would be used?_____

15. A _____ is an attempt to sample the whole population.

16. The _____ _____ is a psychological effect to a treatment in which the subject has confidence.

IV. Problems

17. Sixty people with the symptoms of the common cold were randomly divided into two groups with one group receiving a new drug and the other receiving aspirin and hot chicken soup. The duration of time until the patient experienced relief from the cold was recorded.
a) What is an experimental unit?
b) What is the treatment?
c) What is the response variable?
d) Are there any possible extraneous variables confounded with the treatment on the response variable?
e) Is there a control group? Is one needed?

18. Of the methods of contact in a sample survey, which
a) is the most reliable?
b) has the lowest response rate?
c) is the most often used?
d) is the most expensive to conduct?

19. A real estate developer is studying the possibility of a new convention center in the city. A survey on the issue was sent to 300 randomly selected residents. Describe
a) the population of interest
b) the sample
c) a variable of interest.

20. Suppose a member of the student government association at university X asks every 10th person in the cafeteria line what they think of the food.
a) Is this a survey or experiment?
b) What sampling procedure is used?
c) What population does this sample generalize to?
d) Name two variables of interest.

21. In the personal interview, how might response rate be increased?

22. A county agent randomly selects 50 bee hives in his county to study a disease that affects productivity.
a) What is the population of interest?
b) Would he conduct a survey or observational study?

Answers to Practice Test 1

1. F   2. F   3. T   4. T   5. F   6. F   7. T

8. c   9. c   10. a

11. frame   12. bias   13. randomization

14. stratified   15. census   16. placebo effect

17. a) a person with symptoms of a common cold
    b) drug or aspirin w/chicken soup
    c) duration of time until the patient experienced relief from the cold
    d) yes, severity of cold, age of subject, general health of subject
    e) No, No

18. a) personal interview
    b) mailed questionnaire
    c) telephone interview
    d) personal interview

19. a) residents of the city
    b) the 300 randomly selected residents
    c) opinion concerning the convention center, family income.

20. a) survey
    b) systematic
    c) students at university that eat in the cafeteria
    d) gender, opinion of food

21. Properly trained interviewers and properly worded questions.

22. a) All bee hives in the county.
    b) Observational study.

Practice Test 2

I. True/False

1. The Nielsen ratings are not very accurate because the sample size is so small (around 1500).

2. Because of time and cost, a census should never be conducted.

3. A simple random sample will always give a representative sample.

4. An extremely large sample size will always guarantee a representative sample.

5. A mailed survey is less reliable than a personal interview.

6. A cluster random sample is obtained by dividing the population into groups and then sampling from each group.

7. The control group does not get a treatment.

II. Multiple Choice

8. Catholics are more likely than white Protestants to belong to the working class rather than to the middle class. Working class persons are usually Democrats, while members of the middle class are more often Republicans. So in examining why a person is a Republican, the effects of religion and social class are
(a) confounded.
(b) biased.
(c) independent.
(d) response variables.

9. Immediately before an election, the Gallup Poll interviews about 4000 people instead of the usual 1800. The purpose of this is
(a) to remove bias from their results.
(b) to allow use of a SRS.
(c) to be certain they can predict the election.
(d) to increase the precision of their results.

III. Fill in the blank

10. A _____ experiment is one in which the subjects and those who carry out the experiment are ignorant of who receives the treatment and who does not.

11. A _____ is a subset of the population used to gain information about the whole.

12. A statistician wishes to make a generalization about the _____.

13. A _____ assigns a value to each unit in the population.

14. What sampling technique would be best for surveying the households in a large city?
_____

15. A _____ random sample of size n is a sample in which each sample of size n has the
same chance of being chosen.

16. If a procedure consistently gives low values we say the procedure is _____.

IV. Problems

17. A newspaper article reported the number of farmers in each state who went bankrupt in 1984. For
example, Texas has 428 and North Carolina had 230 go out of business. Is this a meaningful measure
to compare the number of bankrupt farmers in the various states? Can you think of a more valid
measure for comparison?

18. Identify the sampling technique used in the following:
a) An interviewer selects a person every 10 minutes as they leave the student union building.
b) Twenty classes are randomly selected across campus and each student in the class is given a survey
to complete.
c) Samples of size 100, 80, 70, and 60 are randomly selected from the freshman, sophomore, junior,
and senior classes respectively.

19. Statistically speaking, what justification do experimenters have for giving overdoses to animals in
order to study the effects of a treatment?

20. A university is interested in the success their placement office has in placing graduates in their
        chosen fields. A survey concerning the issue was sent to 500 of their past graduates. Describe
a) the population of interest
b) the sample
c) two variables of interest
d) a typical observation

21. A cosmetic company wants to market a new product nationwide. To study consumer preference sample
products were sent to all households in 5 randomly selected cities.
a) What is the population of interest?
b) What sampling technique is used? Describe the sample.
c) Give two variables that would be measured in this study.

Answers to Practice Test 2

1. F   2. F   3. F   4. F   5. T   6. F   7. T   8. a   9. d

10. double-blind   11. sample   12. population

13. variable   14. cluster   15. simple   16. biased

17. No, because the total number of farms is so different in the two states. A better measure is the proportion of bankrupt farms relative to the total number of farms in the state.

18. a) systematic
    b) cluster
    c) stratified

19. The overdose over a short period of time should simulate the long term effect of the treatment.

20. a) Graduates of the university.
    b) The 500 past graduates.
    c) Field of employment.
    d) One graduates field of employment.

21. a) Consumers of cosmetic products.
    b) Cluster sampling.
    c) Household income and how well they liked the product.

2

Organizing And Summarizing Univariate Data

A univariate data set is such that one measurement (one variable) is taken on each subject. For example, you could measure the grade point average (GPA) of a random sample of sophomores from your school. This chapter gives details on how to organize and summarize univariate data to make it more comprehensible.

Important issues—Chapter 2

- A data set consist of a listing of the observed values of one or more variables. A data set with one variable listed is called *univariate*. When two variables are measured on each subject the data set is called *bivariate*. When several variables are measured on each subject the data set is called *multivariate*.

- Variables are classified as either *numerical* or *categorical*. You must be able to distinguish between the two types of variables. Numerical variables are further classified as *discrete* or *continuous*. You must also be able to distinguish between a discrete and continuous variable.

- Data recorded in time are called *time series* data. You must be able to summarize the data by constructing a time series graph.

- Categorical data can be listed in a table called a *frequency table*. The *relative frequency* for a category is obtained by dividing the frequency of the category by the total frequency. You must be able to formulate the data in a frequency table and/or a relative frequency table and illustrate it with a *bar graph* or *pie chart*

- Numerical data can be displayed in a *stem and leaf plot*. Given any set of data you should be able to construct an informative stem and leaf plot. The data are left intact with a stem and leaf plot; however, in some cases a *histogram* or a *frequency polygon* is more appropriate. You must be able to draw a histogram from a stem and leaf plot and construct a frequency polygon from a histogram.

- Populations and samples are characterized by summary measures. A *parameter* is a summary measure associated with a population. A *statistic* is a summary measure associated with a sample. In a given situation you will be asked to decide if a numerical quantity is a parameter or a statistic. If it is a statistic you will also be asked to compute its numerical value. The stem and leaf plot is used to organize sample data so that the values of various statistics can be calculated more easily.

- Summary measures are often classified as *measures of location* and *measures of variability*. Among the measures of location are those that measure the center of the distribution. They include the *mean*, the *trimmed mean*, and the *median*. Other measures of location are the *quartiles*, the *letter values*, and the *high* and the *low*. From these, other measures of center can be found. They are called midsummaries and include the *midQ* and the *midrange*. The high, low, quartiles and median are summarized in a 5-number summary diagram. For any given data set you should be able to compute any of the summary measures described in this chapter.

- Measures of variability include the *range*, the *standard deviation*, the *Q-spread*. You should know the definition of each and be able to compute the numerical value of any measure of variability described in this chapter.

- Standard deviations can be interpreted with the *empirical rule* and *Chebyshev's rule*. The empirical rule applies to bell-shaped distributions and Chebyshev's rule may be applied to any shaped distribution. Given any data set you should be able to give an interpretation of standard deviation using either the empirical rule or Chebyshev's rule.

- The *boxplot* is a graphical tool to picture the data and detect possible *outliers* in the data. Know how to construct a boxplot by hand and with a computer.

- You need to be able to classify the shape of a distribution. A *skewed* distribution is a distribution with one tail significantly longer than the other. It is said to be skewed in the direction of the long tail. A *symmetric short-tailed* distribution has tails that drop off more rapidly that the tails of a normal curve. A *symmetric long-tailed* distribution has tails longer than the tails of a normal curve. A *multi-modal* distribution has more than one peak. Typically, the *bimodal* distribution with its two peaks is more commonly observed in real data. The occurrence of more than one mode in a data set may be an indication of a nonhomogeneous factor in the data. If at all possible, the factor should be identified.

Glossary

A **univariate data set** is a data set in which one measurement (variable) has been made on each experimental unit.

A **bivariate data set** is a data set in which two measurements (variables) have been made on each experimental unit.

A **multivariate data set** is a data set in which several measurements (variables) have been made on each experimental unit.

A **numerical variable** (also called a quantitative or measurement variable) is a variable whose values are numbers obtained by a count or measurement.

A **discrete variable** is a numerical variable that can assume a finite number or at most a countable infinite number of values. (Countable means you can associate the values with the counting numbers 1, 2, 3, --- ; that is, the values can be counted.)

A **continuous variable** is a numerical variable that can assume an infinite number of values associated with the numbers on a line interval.

A **categorical variable** (also called a qualitative variable) is a variable whose values are classifications or categories.

A table listing the different categories of categorical data and the corresponding frequencies with which they occur is called a **frequency table.**

A **bar graph** is a picture consisting of horizontal and vertical axes with rectangles representing the frequency (or relative frequency) of the data. The values of the variable are listed along one axis and the frequency (or relative frequency) along the other.

The **stem and leaf plot** is a quick and easy way to display numerical data. It is extremely useful for arranging the observations from smallest to largest so that specific locations in the data set can be found.

The **histogram** is an alternate way of graphically presenting numerical data. It can be constructed from the stem and leaf plot. Each stem of a stem and leaf plot defines an interval of values called a **class**. The **class limits** are the smallest and largest possible values for the leaves of that stem.

Once the class limits are determined the data can be formulated in a **grouped frequency table** (grouped in the sense that the data has been grouped into the various classes) The **class boundaries** are obtained by lowering the lower class limits by .5 and raising the upper class limits by .5.

The **frequency polygon** is another useful tool for describing numerical data. It is a graph of connecting line segments that correspond to the frequencies of the various classes.

A **parameter** is a numerical summary measure of a population distribution.

A s**tatistic** is a numerical quantity calculated from the observations in a sample.

The **population mean**, denoted by the Greek letter μ, is the numerical value that locates the *balance point*, also called the center of mass, of the population distribution. If a sample consists of observations $y_1, y_2, y_3, ---, y_n$ then the **sample mean** is $\bar{y} = (y_1+y_2+---+y_n)/n = \Sigma y_i/n$.

A **p% trimmed population mean** denoted by μ_T is obtained by trimming p% of the distribution off both ends and finding the mean of the remaining distribution. A **p% trimmed sample mean** denoted by $\bar{y}_T$ is found by calculating p% of n and then trim that many observations off both ends of the ordered data and then calculate the mean of the remaining observations. If p% of n is not a whole number *round up* to the next whole number.

The **population median**, denoted by θ, is the numerical value that divides the population distribution in half. If the sample observations y_1, y_2, ---, y_n are arranged in order from smallest to largest, the **sample median**, denoted by M, is the middle observation if n is odd, or the average of the two middle observations if n is even. In either case, the median is located at the position $(n + 1)/2$ in the ordered data set.

The **first quartile**, denoted by θ_1, is the numerical value that divides the lower half of the population distribution in half. The **third quartile**, denoted by θ_3, is the value that divides the upper half of the population distribution in half. The **first** and **third sample quartiles**, Q_1 and Q_3, are similarly defined for samples. The median is the **second quartile**, Q_2

The **range** is the highest measurement minus the lowest measurement, H - L.

The **population variance**, σ^2, is the average squared distance of all measurements from the population mean. The **sample variance**, s^2, is an average squared distance of the sample values from the sample mean. It is calculated with the formula:

$$s^2 = \frac{\Sigma(y_i - \overline{y})^2}{n - 1}$$

The expression in the numerator is referred to as a **sum of squares** which measures the *total deviation* of all the data. It is denoted as $\quad SS = \Sigma(y_i - \overline{y})^2$

In either case, population or sample, the positive square root of the variance is called the **standard deviation**. The population standard deviation is denoted by σ. The sample standard deviation is

$$s = \sqrt{\frac{\Sigma(y_i - \overline{y})^2}{n - 1}}$$

The **z-score** corresponding to a particular observation is $z = (observation - mean)/standard\ deviation$ and gives the number of standard deviations the observation is from the mean.

The **Q-spread** is the distance between the first and third sample quartiles, $Q_3 - Q_1$. The corresponding population **q-spread** is similarly defined using the population quartiles in place of the sample quartiles.

A distribution is said to be **symmetric** if the scores below the center of the distribution are a mirror image of the scores above the center.

A frequency distribution whose *left* tail is longer than the right is called **skewed left**. If the *right* tail is longer than the left then it is called **skewed right**.

A symmetrical distribution is called **short-tailed** if the tails of its frequency curve drop off more rapidly than the tails of a normal curve. It is called **long-tailed** if the tails of its frequency curve are somewhat longer than the tails of a normal curve.

A distribution is called **multi-modal** if its frequency curve has two or more peaks. A **mode** is the value associated with a peak.

Practice Test 1

I. Multiple Choice

1. A political scientist records the income (in dollars) and the political affiliation of 400 urban wage earners. The number of variables measured by the political scientist is
a) one.
b) two.
c) three.
d) 400.

2. In problem #1, political affiliation is
a) a categorical variable.
b) a continuous numerical variable.
c) a discrete numerical variable.
d) none of the above.

3. The variable--Number of blue whales in the ocean-- is
a) categorical.
b) categorical and continuous.
c) numerical and continuous.
d) numerical and discrete.

4. The variable--Brand of video tape-- is
a) categorical.
b) categorical and discrete.
c) numerical and continuous.
d) numerical and discrete.

5. The variable--GPA of students who participate in sports-- is
a) continuous.
b) discrete.
c) neither continuous nor discrete it is categorical.

6. The variable--Number of unemployed Hispanics in the city-- is
a) continuous.
b) discrete.
c) neither continuous nor discrete it is qualitative.

7. The variable--Corporations with assets over $10 million-- is
a) continuous.
b) discrete.
c) neither continuous nor discrete it is qualitative.

8. A poll of 2500 adults attending a professional football game were asked if they thought that the instant replay improves officiating. Thirty-three percent of the males and 41% of the females agreed that it improved the officiating of the game. In this situation, the 41% is

a) a parameter   b) a statistic   c) a sample   d) a variable

9. In Exercise 1 the gender of the respondent is

a) a parameter   b) a statistic   c) a sample   d) a variable

10. Which of the following measure variability in a data set?

a) sample mean   b) sample median   c) sample midrange   d) sample Q-spread

11. Which of the following are resistent to the influence of outliers? (An outlier is an extremely large or extremely small score in comparison to the other scores.)

a) sample mean   b) sample median   c) sample midrange   d) sample range

II. Fill in the blank

12. Time series graphs should always have time on the _____ axis.

13. Classify the following as either categorical or numerical. If numerical further classify as discrete or continuous.

Variable	Categorical or numerical	Discrete or Continuous
number of books read by 3rd graders	_____	_____
the final grade students receive in this course	_____	_____
duration of a tennis match	_____	_____

43

courses offered by a
university

_____   _____

number of penalties
in a football game

_____   _____

title of books read
by 3rd graders

_____   _____

the 10 most favorite books
chosen by 3rd graders

_____   _____

house paint color

_____   _____

cost of a haircut
in your city

_____   _____

rating of movies

_____   _____

millionaires living
in Charlotte, NC

_____   _____

III. Problems

14. What is the relationship between the mean and the median in a skewed right distribution?

15. Identify the following as a parameter or a statistic.
a) The crime rate for Charlotte, NC last year was **17.6** per 100,000.
b) The average reaction time of 30 senior citizens taken on an instrument at the drivers license bureau is **84.63**.
c) The lowest agressive tendency score by 28 teenagers was **11**.
d) The mortality rate for a certain disease is **30%**.
e) The middle salary of 25 new PH.D. psychologists was **$25,200**.

16. Of the 254 reported deaths due to infective and parasitic diseases in the state of Hawaii in 1994, 35 were in Hawaii, 183 in Honolulu, 13 in Kauai, and 23 in Maui. Organize these data in a frequency table. Draw a bar graph of the data.

17. What is wrong with the following pie chart?

Pie chart

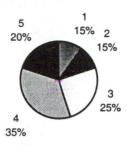

5
20%

1
15%

2
15%

3
25%

4
35%

18. Following are test grades on the first test given in an introductory chemistry class:

78, 73, 85, 66, 48, 79, 49, 88, 69, 71, 78, 84, 89, 53, 51, 62, 64, 80, 66, 91, 93, 84, 94, 35

a) Construct a stem and leaf plot for the data.
b) Find the mean, median and 5% trimmed mean.
c) Find the standard deviation.
d) Complete a 5-number summary diagram.
e) Describe the general shape of the distribution.

19. Suppose there are 3 machines to bottle a brand of 12 oz beer bottles. Suppose samples of size 8 are selected from each machine and the contents of the bottle measured. From the following data evaluate the machines with side-by-side boxplots. (see Example 2.26 pg 132)

Contents in ounces
Machine A 11.32, 11.67, 11.91, 11.74, 11.82, 11.75, 11.88, 11.41
Machine B 11.92, 11.96, 12.00, 11.95, 11.87, 11.91, 11.96, 11.88
Machine C 11.25, 11.28, 11.27, 11.23, 11.24, 11.30, 11.35, 11.29

20. From the following data that has been organized in a stem and leaf plot complete a 5-number summary display to include the midsummaries and spreads.

```
 3 | 2 4 1
 4 | 3 0 6 3 8
 5 | 8 1 6 3 9 6
 6 | 4 9 2 6 5
 7 | 9 2 3 0
 8 | 3 5
 9 | 2
10 | 0
```

21. A researcher is interested in studying the length of marriages that end in divorce. Following are the number of years of marriage obtained from a random sample of divorce records:

1.7, 2.6, 1.9, 6.5, 2.8, 3.1, 7.2, 15.8, 2.5, 3.4, 1.3, 0.2, 4.5, 2.4, 35.0, 7.5, 3.2, 2.1, 1.8, 5.0, 6.6, 4.2, 3.1, 12.8, 3.5, 2.9
a) Construct a stem and leaf plot.
b) Construct a 5-number summary display to include the midsummaries and spreads.
c) Calculate the sample mean and standard deviation.

Answers to Practice Test 1

1. b　　2. a.　　3. d　　4. a　　5. a　　6. b　　7. c　　8. b　　9. d　　10. d　　11. b

12. horizontal

13.　　Numerical, Discrete
　　　　Categorical
　　　　Numerical, Continuous
　　　　Categorical
　　　　Numerical, Discrete
　　　　Categorical
　　　　Categorical
　　　　Categorical
　　　　Numerical, Continuous
　　　　Categorical
　　　　Numerical, Discrete

14. The mean is greater than the median in a skewed right distribution

15. a. statistic　b. statistic　c. statistic　d. parameter　　e. statistic

16.

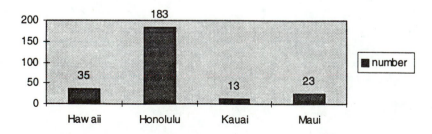

17. The percents do not add to 100%.

18. a.

```
3 | 5
4 | 8 9
5 | 1 3
6 | 2 4 6 6 9
7 | 1 3 8 8 9
8 | 0 4 4 5 8 9
9 | 1 3 4
```

b. Mean =72.08, Median =75.50, TrMean = 72.77

c. StDev = 16.15

d.
Low	Q1	M	Q3	High
35	63	75.5	84.5	94

e. The distribution is skewed left

19.

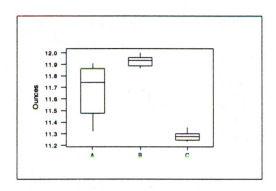

20. **Letter Value Display**

	DEPTH	LOWER	UPPER	MID	SPREAD
N=	27				
M	14.0	59.000		59.000	
H	7.5	47.000	71.000	59.000	24.000
	1	31.000	100.000	65.500	69.000

21. a. **Stem-and-leaf of years   N = 26**
    Leaf Unit = 1.0

```
(18)   0 011112222223333344
  8    0 56677
  3    1 2
  2    1 5
  1    2
  1    2
  1    3
  1    3 5
```

b. **Letter Value Display**

	DEPTH	LOWER	UPPER	MID	SPREAD
N=	26				
M	13.5	3.150		3.150	
H	7.0	2.400	6.500	4.450	4.100
	1	0.200	35.000	17.600	34.800

47

```
c.  Descriptive Statistics

Variable        N     Mean   Median   TrMean   StDev   SEMean
years          26     5.52     3.15     4.52     6.95    1.36
```

Practice Test 2

I. Multiple Choice

1. In order to be correct, a line graph showing the change over time in number of abortions in the U. S. must have
a) number of abortions on the horizontal scale.
b) time on the horizontal scale.
c) either variable on the horizontal scale, as long as equal scale intervals are used.
d) bars of equal width.

2. The variable--Amount of oil lost at sea-- is
a) categorical.
b) categorical and continuous.
c) numerical and continuous.
d) numerical and discrete.

3. The variable--Outcome of a horse race-- is
a) categorical.
b) categorical and discrete.
c) numerical and continuous.
d) numerical and discrete.

4. The variable--Number of ties on a rack-- is
a) categorical.
b) categorical and discrete.
c) numerical and continuous.
d) numerical and discrete.

5. The variable--Number of defective computer parts-- is
a) categorical.
b) categorical and continuous.
c) numerical and continuous.
d) numerical and discrete.

6. The variable--Production cost to produce a product-- is
a) continuous.
b) discrete.
c) neither continuous nor discrete it is categorical.

7. The variable--Brand of computer-- is

a) continuous.
b) discrete.
c) neither continuous nor discrete it is categorical.

8. The variable--Number of students who own cars-- is
a) continuous.
b) discrete.
c) neither continuous nor discrete it is categorical.

9. The variable--Choice of TV program-- is
a) continuous.
b) discrete.
c) neither continuous nor discrete it is categorical.

10. It has been reported that 40% of all restaurant food is consumed outside the restaurant. The 40% is
a) a parameter
b) a statistic
c) a sample
d) a variable

Circle the statements that are true:
11. Regarding summary statistics
a) the mean and standard deviation are measures of location
b) the range and standard deviation are measures of location
c) the mean and median are measures of location
d) the median and range are measures of location
e) the mean and standard deviation are measures of variability.
f) the range and standard deviation are measures of variability.
g) the mean and median are measures of variability.
h) the median and range are measures of variability.
i) the median and IQR are resistant to outliers.

II. Fill in the blank

12. Classify the following as either categorical or numerical. If numerical further classify as discrete or continuous.

Variable	Categorical or numerical	Discrete or Continuous
quality of building material	_____	_____
batting average of hall of fame players	_____	_____
duration of an opera	_____	_____

color of autos in
a parking lot   _____   _____

California achievement
test scores   _____   _____

rating of pillows
for softness   _____   _____

type golf ball used
by pros   _____   _____

number of attempts to
complete a task   _____   _____

movies playing in town
on the weekend   _____   _____

age of freshmen senators _____   _____

III. Problems

13. What is the difference in a parameter and a statistic?

14. Data from the U.S. Census Bureau indicates that last year 29% of all households had 1 wage earner, 42% had 2 wage earners, 10% had 3 wage earners, 4% had four or more, and 15% had no wage earner. Organize the data in a relative frequency table. Draw a bar graph of the data.

15. Statistics from the National Fire Protection Association show that from 1978 through 1985 the number of people who have died yearly in fires is

   6015, 5500, 5250, 5400, 4850, 4600, 4075, and 4938.

a) Illustrate this data in a time series graph.
b) Illustrate the same data in a bar graph.

16 . Following are test grades on the first test given in an introductory English class:

   65, 73, 85, 66, 43, 79, 46, 86, 63, 71, 78, 74, 89, 54, 52, 61, 63, 80, 68, 77, 95, 82, 28, 70

a) Construct a stem and leaf plot for the data.
b) Find the mean, median and 5% trimmed mean.
c) Find the standard deviation.
d) Complete a 5-number summary diagram.
e) Describe the general shape of the distribution.

17. Construct a grouped frequency table to include class limits and class boundaries for the data in exercise 15. From the grouped frequency table construct a histogram.

18. Following are the average yards per reception for the pass receivers on a professional football team.

    10.4, 16.5, 9.4, 11.4, 11.4, 22.3, 10.2, 7.0, 6.5, 19.0, 18.0, 11.0

    a) Organize the data in a stem and leaf plot.
    b) Find the mean, median, 5% trimmed mean, and standard deviation.

19. A researcher for a long-distance telephone company is interested in studying the length of a typical long-distance phone call. Following are the results (recorded in minutes) of a random sample of 35 calls:

    12.8, 3.5, 2.9, 9.4, 8.7, 3.5, 4.8, 7.7. 5.9, 6.2, 2.8, 4.7, 1.7, 2.6, 1.9, 6.5, 2.8, 3.1, 7.2, 15.8, 2.5, 3.4, 1.3, 0.2, 4.5, 2.4, 35.0, 7.5, 3.2, 2.1, 1.8, 5.0, 6.6, 4.2, 3.1,

a) Construct a 5-number summay display to include the midsummaries and spreads.
b) Using your calculator, calculate the sample mean, 5% trimmed mean, and the standard deviation.

20. Following are the batting averages of the active players on the Baltimore Orioles.

    .329 .325 .282 .274 .271 .266 .265 .264 .259
    .258 .255 .226 .218 .216 .195 .187 .132 .111

a) Organize the data in a stem and leaf plot.
b) Find the mean and 5% trimmed mean.

Answers to Practice Test 2

1. b   2. c   3. a   4. d   5. d   6. a   7. c   8. b   9. c   10. a   11. c, f, and i

12.   Categorical
        Numerical, Continuous
        Numerical, Continuous
        Categorical
        Numerical, Continuous
        Categorical
        Categorical
        Numerical, Discrete
        Categorical
        Numerical, Continuous

13. A parameter is a numerical characteristic of a population and a statistic is a numerical characteristic of a sample.

14.
number of wage earners	percent
only 1	29
exactly 2	42
exactly 3	10
four or more	4
none	15

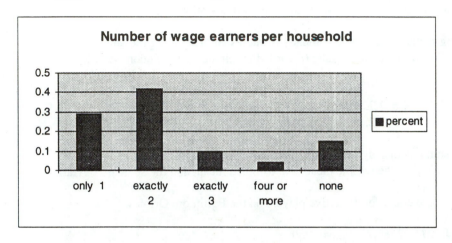

15.   a

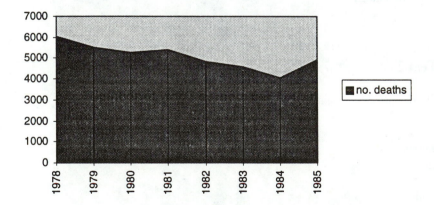

52

b.

Number of deaths due to house fires

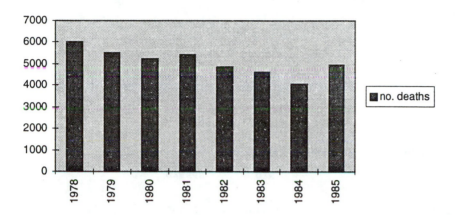

16. **Stem-and-leaf of C1   N = 24**
Leaf Unit = 1.0

```
 1      2 8
 1      3
 3      4 36
 5      5 24
11      6 133568
(7)     7 0134789
 6      8 02569
 1      9 5
```

b. Mean =68.67, Median =70.50, TrMean = 69.32

c. StDev = 15.83

d. Low   Q1   M   Q3   High
   28.00   62   70.5   79.5   95.00

e. The distribution is skewed left

17.

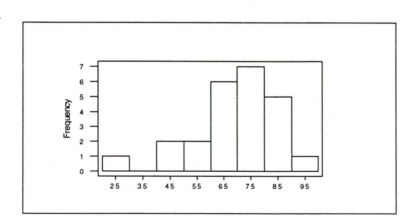

18. a. **Stem-and-leaf of C1   N = 12**
    Leaf Unit = 1.0

```
    2      0 67
    3      0 9
   (5)     1 00111
    4      1
    4      1
    4      1 6
    3      1 89
    1      2
    1      2 2
```

   b. | Variable | N | Mean | Median | TrMean | StDev |
      |---|---|---|---|---|---|
      | C1 | 12 | 12.76 | 11.20 | 12.43 | 4.99 |

19. a.   **Letter Value Display**

	DEPTH	LOWER	UPPER	MID	SPREAD
N=	35				
M	18.0		3.500	3.500	
H	9.5	2.700	6.550	4.625	3.850
	1	0.200	35.000	17.600	34.800

   b. | Variable | N | Mean | Median | TrMean | StDev |
      |---|---|---|---|---|---|
      | minutes | 35 | 5.64 | 3.50 | 4.68 | 6.06 |

20. a. **Stem-and-leaf of batting   N = 18**
Leaf Unit = 0.010

```
    1      1 1
    2      1 3
    2      1
    2      1
    4      1 89
    6      2 11
    7      2 2
   (3)     2 555
    8      2 66677
    3      2 8
    2      3
    2      3 22
```

   b. | Variable | N | Mean | Median | TrMean |
      |---|---|---|---|---|
      | batting | 18 | 0.2407 | 0.2585 | 0.2433 |

3

Bivariate Data—
Studying Relations Between Variables

This chapter addresses relationships between two variables. Recall that in Chapter 2 variables were classified as either categorical or numerical. We organize and summarize the relationship between two categorical variables, between a categorical and a numerical variable, and finally between two numerical variables.

Important issues—Chapter 3

- The relationship between categorical variables is displayed in a two-way frequency table called a *contingency table*. An important concern is whether or not the variables are independent.

- The association between categorical variables is also illustrated in a *segmented bar graph*.

- Side-by-side boxplots can be used to display bivariate data in which one variable is numerical and the other is categorical.

- The relationship between numerical variables is displayed in a *scatterplot*. It is useful in detecting trends in the data and revealing an association between the two variables.

- The *correlation coefficient* is used to numerically measure the degree of association between numerical variables. We considered two measures of correlation—*Pearson's correlation coefficient, r,* and *Spearman's rank correlation, r_s.*

- The basic properties of Peason's correlation are:

  * The value of r measures the *linear* relationship between x and y and will always be between -1 and +1.
  * The closer it is to either -1 or +1 the stronger the linear relationship between x and y. In fact, points that fall exactly on a straight line have a correlation of +1 if the line has positive slope and -1 if the line has negative slope.
  * If r is 0 then x and y are not linearly related. They may be related but the relationship would not be that of a straight line.
  * The value of r is not changed when the units of measurement are changed.

- Unlike Peason's correlation, Spearman's is *resistant* to the influence of outliers. It treats all observations equally, so that outliers have no more effect on the outcome than any other observation. Additionally, Spearman's correlation identifies both linear and nonlinear relationships.

- An observation pair in a scatterplot that is far removed from the overall pattern of the remaining data, is called a *bivariate outlier*. They often make it difficult to construct a scatterplot and may have a dramatic effect upon the value of the correlation coefficient.

- Occasionally a simple transformation of the data, such as taking a logarithm or square root of each observation, can reveal relationships between variables that might otherwise go undetected.

- A *lurking variable* is a third variable that may be the cause of an observed correlation between the two variables.

- Important relationships are positive and negative linear trends. They are numerically described with the least squares regression line.

- In the regression setting we distinguish between the *response variable*, y, and the *predictor variable*, x.

- The equation of the line that best fits the data, called the *regression line*, is given by $\hat{y} = b_0 + b_1 x$.

- Once found, the regression line can be used to predict new values of the response variable from different values of the predictor variable. Using the regression line to predict outside the range of the observed data is called *extrapolation*.

- Having found the least squares regression line, we next ask how well does it fit the data? Does the line effectively describe the relationship between x and y? To evaluate, the *coefficient of determination*, r^2 is used to measures the proportion of the variability in y that is explained by x through a linear relationship.

- The *residuals*, $e_i = y_i - \hat{y}_i$, measures the discrepancy between the line and the observed data, a plot of the residuals versus the predictor variable will graphically illustrate the deficiencies of the line as a model of the x-y relationship.

- Outliers, if present in the data, can have an undue effect on the regression line.

- The *resistant line*, an alternative to the least squares regression line, is used in an effort to prevent outliers from distorting the analysis.

Glossary

Given the n observations (x_1,y_1), (x_2,y_2), ..., (x_n,y_n) of the variables x and y, the **covariance between x and y** is $\quad s_{xy} = \Sigma(x_i - \overline{x})(y_i - \overline{y})/(n-1)$

The **Pearson correlation coefficient**, denoted by r, is given by

$$r = \frac{s_{xy}}{s_x\, s_y} = \frac{\Sigma(x_i - \overline{x})(y_i - \overline{y})}{\sqrt{\Sigma(x_i - \overline{x})^2 \Sigma(y_i - \overline{y})^2}}$$

The **population correlation coefficient** is denoted by the Greek letter ρ, (rho).

A measure of correlation for categorical data that can be ranked is given by **Spearman's rank correlation coefficient**, r_s. Given the sample $(x_1,y_1),(x_2,y_2),$ ---, (x_n,y_n), ranks of 1 to n are assigned to the x's and to the y's separately. For tied observations, the average of the ranks that would have been assigned had there been no ties is given to each tied observation. The formula for r_s is the same as the formula for Pearson's correlation coefficient except that it is applied to the ranks of the data.

A **response variable,** also called a **dependent variable,** is the variable we wish to predict or describe based on the values of another variable. The **predictor variable,** also called the **independent variable,** is the variable that is used to explain the response variable. In the context of regression, the response variable will be labeled **y** and the predictor variable will be labeled **x.**

From the data point (x_i,y_i) the observed value of y is y_i and the **predicted value of y** is obtained by the equation $\hat{y}_i = b_0 + b_1 x_i$

The error of the prediction (also called the *residual*) is the difference between the actual y_i and the predicted $\hat{y}_i$.

The **residual** associated with the data point (x_i,y_i) is $\quad e_i = y_i - \hat{y}_i$

The **sum of squares due to error** (also called the residual sum of squares) is given by $SSE = \Sigma(y_i - \hat{y}_i)^2$

Least Squares Regression line is $\quad \hat{y} = b_0 + b_1 x$

where $\quad b_1 = \dfrac{s_{xy}}{s_x^2} = \dfrac{\Sigma(x_i - \overline{x})(y_i - \overline{y})}{\Sigma(x_i - \overline{x})^2}$

and $\quad b_0 = \overline{y} - b_1 \overline{x}$

The **coefficient of determination, r^2,** is the percent of the variability in the dependent variable that is explained by the independent variable.

The equation of the **resistant line** can be written as $\hat{y}_R = b_0 + b_1 x$ where b_1 is the slope and is given by

$$b_1 = \frac{y_R - y_L}{x_R - x_L}$$

and b_0 is the y-intercept given by $\quad b_0 = 1/3[(y_L + y_M + y_R) - b_1(x_L + x_M + x_R)]$.

Practice Test 1

I. Multiple Choice

1. Pearson's correlation measures
a) variability
b) association
c) spread
d) location

2. Regarding the two measures of correlation that are given
a) Pearson's correlation is resistant to outliers.
b) Spearman's correlation is resistant to outliers.
c) Spearman's correlation cannot be applied to continuous data.
d) Pearson's correlation cannot be applied to discrete data.

3. The expression $(y_i - \bar{y})$ in the formula for s^2 is called

a) a coefficient of determination.
b) a standard deviation.
c) a deviation from the mean.
d) an averaged deviation.

4. The resistant line is
a) influenced by outliers.
b) not influenced by outliers.
c) found by the method of least squares.
d) based on averages.

5. Suppose the least squares regression line is given by $y = 1.6 - 3.4x$. The predicted y when x = 5 is
a) -15.4

b) 18.6

c) 4.6

d) 17

6. Given the regression line in Exercise 5, the residual for the data point (5, -12) is

a) -3.4

b) 3.4

c) 7

d) -7

7. The relationship between two categorical variables is displayed in

a) a scatterplot

b) a boxplot

c) a contingency table

d) none of the above

8. The relationship between two numerical variables is displayed in

a) a scatterplot

b) a boxplot

c) a contingency table

d) none of the above

II. Problems

9. The state highway patrol desires to investigate the issue of drivers who wear seat belts. A random selection of 400 drivers revealed that 97 were less than 26 years old, 106 were between 26 and 35, 97 were between 36 and 55, and 100 were 56 or older. Of the 97 less than 26, only 23 wear their seat belts on a regular basis, 41 of those in the 26 to 35 age group wear seat belts, 68 of those in the 36 to 55 group wear seat belts, and 26 of those over 55 wear seat belts.

a) Organize the data in a two-way contingency table.

b) What percent of those wearing seat belts are less than 26 years old?

c) What percent of those less than 26 years of age wear seat belts?

d) What percent wear seat belts?

e) What percent of those over 55 wear seat belts?

f) What percent are over 55 and wear seat belts?

10. Construct a bar graph of the data in Exercise 9.

11. Calculate Pearson's correlation coefficient for the following data:

```
x |  16   22   17   34   25   18
--|------------------------------
y |   3    6    2   11    8    4
```

12. Calculate Spearman's correlation coefficient for the data given in the previous exercise.

13. The following scatterplots represent the relationship between the AIDS rates per 100,000 residents for D. C. and the 50 states for 1994 and 1995. The first scatterplot includes the 94 and 95 rates for D. C. and the second does not.
   a)   What do we call the data point associated with D. C.?
   b)   What effect will it have on the correlation if we leave it in the data set?
   c)   Is the correlation high or low?
   d)   Describe the relationship between the 1994 AIDS rate and the 1995 AIDS rate.

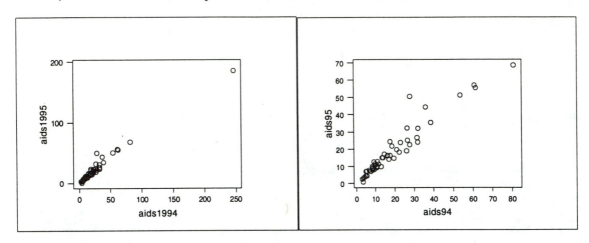

14. Construct the least squares regression line from the following data:

```
x  |   12   23   16   22   31   18   25   29
---|-----------------------------------------
y  |    3   16    8   19   24   10   18   21
```

15. Construct the resistant line from the data in Exercise 13.

16. The visual acuity of 9 subjects is tested under different doses of a drug. Do the following data suggest a linear relation between the drug dose and the visual acuity? Find the regression equation relating the two variables and calculate the correlation.

Subject	1	2	3	4	5	6	7	8	9
Drug dose	5	5	5	10	10	10	15	15	15
Visual acuity	.36	.41	.55	.33	.28	.25	.17	.14	.20

Answers to Practice Test 1

1. b   2. b   3. c   4. b   5. a   6. a   7. c   8. a
9. a.

	less than 26	26 to 35	36 to 55	over 55	Total
seatbelts	23	41	68	26	158
no belts	74	65	29	74	242
Total	97	106	97	100	400

60

b. 14.56% c. 23.7% d. 39.5% e. 26% f. 6.5%

10.

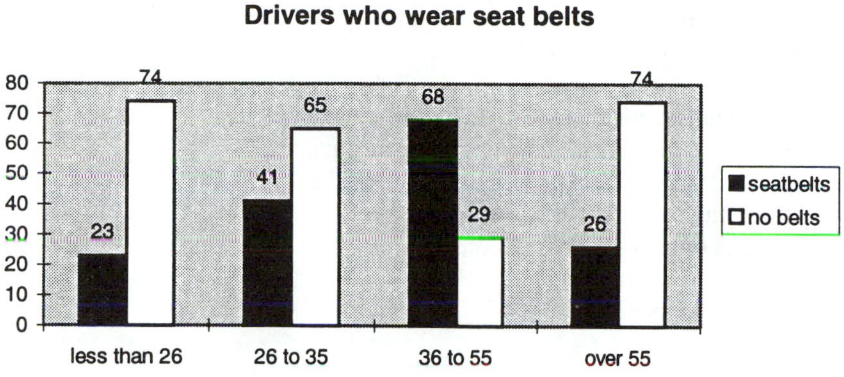
Drivers who wear seat belts
80
70
60
50
40
30
20
10
0
74
65
68
74
23
41
29
26
seatbelts
no belts
less than 26
26 to 35
36 to 55
over 55

11. r = .975 12. r_s = .943

13. a. an outlier b. It will increase the correlation. c. The correlation is high
d. There is a strong linear positive relationship between the 1994 AIDS rate and the 1995 AIDS rate.

14. The regression equation is y = - 8.86 + 1.08 x

15. The resistant equation is y = -7 + x

16. The regression equation is visual = 0.569 - 0.0270 dose. Correlation of dose and visual = -0.902

Practice Test 2

I. Multiple Choice.

1. Which of the following measure association in a data set?
a) sample mean
b) sample median
c) sample midrange
d) none of the above

2. Which of the following are resistent to the influence of outliers?
a) sample mean
b) Pearson's correlation
c) sample midrange
d) Spearman's correlation

3. Lurking variables may have an affect on
a) only Pearson's correlation.
b) only Spearman's correlation.
c) both Pearson's and Spearman's correlations.
d) neither Pearson's nor Spearman's correlation.

4. Regarding the two measures of correlation that are given in this chapter
a) Pearson's correlation identifies both linear and nonlinear relationships.
b) Spearman's correlation identifies both linear and nonlinear relationships.
c) Pearson's correlation identifies only nonlinear relationships.
d) Spearman's correlation identifies only nonlinear relationships.

5. The coefficient of determination is a measure of
a) variability
b) association
c) location
d) none of the above.

6. Suppose the least squares regression line is given by $y = 5.6 + 2.5x$. The predicted y when x = -2 is
a) 10.6
b) 0.6
c) 6.1
d) 3.1

7. Given the regression line in Exercise 6, the residual for the data point (-2, 1) is
a) -0.6
b) -0.4
c) +0.6
d) +0.4

8. A large positive correlation between two variables means that
a) as one variable increases the other decreases.
b) large values on the first variable caused large values in the second variable.
c) as one variable increases the other increases.
d) there is a lurking variable causing the association.

II. Problems

9. Clinical depression is a serious condition that may respond to medication. The symptoms include loss of appetite, insomnia, forgetfulness, fatigue, and restlessness. A survey of 1850 adult males and 1900 adult females across the nation found that out of the 1850 males, 350 were at high risk of clinical depression, 425 were at moderate risk and the remaining were not at risk. From the 1900 females it was found that 420 were at high risk, 560 were at moderate risk and the remaining were not at risk of clinical depression.
a) Organize the data in a two-way contingency table.

b) What percent of the females are at high risk?

c) What percent of those at moderate risk are female?

d) What percent of the males are at no risk?

e) What percent are female and high risk?

f) What percent are male and high risk?

10. Construct a bar graph of the data in Exercise 9.

11. Calculate Pearson's correlation coefficient for the following data:

x	6	2	7	4	5	8
y	2	5	0	4	3	1

12. Calculate Spearman's correlation coefficient for the data given in the previous exercise.

13. Give the resistant line for the following data.

x	1.2	2.3	3.5	4.1	4.7	5.8	6.0	7.2	12.4
y	10.8	9.6	10.2	6.5	7.8	6.3	3.8	3.4	2.5

14. In 1991, fifty percent of the 14,850,000 rental properties in the United States were financed with a mortgage. (The following figures are in thousands) Of the 11,285 properties consisting of 1 to 4 units, 5,565 were mortgaged. Of the 557 properties consisting of 5 to 49 units, 387 were mortgaged. Of the 65 properties consisting of 50 or more units, 57 were mortgaged. Of the 1,588 condominiums, 1,059 were mortgaged. And of the 1,355 mobile homes, 276 were mortgaged.

a) Organize the data in a two-way contingency table consisting of the type of property and whether or not the property is mortgaged.

b) What percent of the mortgaged properties are mobile homes?

c) What percent of the mobile homes are mortgaged?

d) What percent of the condominiums are mortgaged?

e) What percent of the mortgaged properties consist of 1 to 4 units?

f) What percent of the 1 to 4 unit properties are mortgaged?

15. Draw a bar graph that illustrates how the mortgaged properties are distributed in Exercise 13.

16. Following are the basal metabolic rate and a condition index for 12 joggers. Find the regression equation relating the two variables and calculate the correlation.

Basal metabolic rate	206	238	224	257	230	236	209	278	267	252	245	210
Condition index	92	84	89	70	81	87	97	45	58	65	79	94

Answers to Practice Test 2

1. d 2. d 3. c 4. b 5. b 6. b   7. d 8. c

9. a.

	high risk	moderate risk	not at risk	Total
male	350	425	1075	1850
female	420	560	920	1900
Total	770	985	1995	3750

b. 14.56%   c. 23.7%   d. 39.5%   e. 26%   f. 6.5%

10.

Risk of Clinical Depression

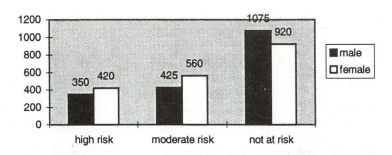

11. $r = -.94$ 12. $r_s = -.943$ 13. The resistant equation is $y = 13.28 - 1.39 x$

14.   a.

	1to4 units	5to49 units	50 or more units	condominiums	mobile homes	Total
mortgage	5565	387	57	1059	276	7344
no mortgage	5720	170	8	529	1079	7506
Total	11285	557	65	1588	1355	14850

b.   3.8%   c. 20.4%   d. 66.7%   e. 75.8%   f. 49.3%

15.

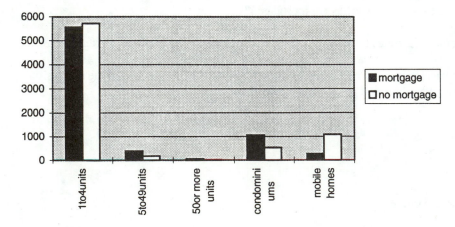

16.   The regression equation is rate = 347 - 1.39 index. Correlation of rate and index = -0.950

4

Probability and Probability Distributions

This chapter introduces the vocabulary and laws of probability and presents problems that illustrate how probabilities are determined. Knowledge of probability and how it applies to statistical analysis is a necessity in almost all fields of study.

Important issues—Chapter 4

- An *experiment* is an activity of making an observation or taking a measurement. The collection of all outcomes associated with the experiment is called the *sample* space. A *tree diagram* will help in listing all the outcomes.

- Any subset of the sample space is called an *event*. The likelihood of the occurrence of an event depends on its *probability*. You must be prepared to discuss the intuitive concepts of probability.

- The probability of an outcome of an experiment can be described as the *relative frequency* with which the outcome will occur if we repeat the experiment a large number of times. If probabilities can be assigned to the outcomes in a sample space, then the probability of any other event A, is the sum of the probabilities that have been assigned to the outcomes that are in event A.

- To assign probabilities to the individual outcomes in a sample space so that probabilities of other events can be determined, two conditions must be satisfied:
  - * The probability of each outcome must be between 0 and 1.
  - * The probabilities of all outcomes in the sample space must sum to 1.

- Most interesting events are formed by taking intersections and unions of other events. *Venn diagrams* are useful for illustrating these *compound events*. Certain laws of probability are used to determine probabilities of compound events.

- The *additive law* is $P(A \text{ or } B) = P(A) + P(B) - P(A \text{ and } B)$

- For *mutually exclusive* events A and B, $P(A \text{ and } B) = 0$, so that $P(A \text{ or } B) = P(A) + P(B)$.

- The *complement law* says that $P(A^c) = 1 - P(A)$.

- Independent events are such that if one event occurs it will not affect the probability of the other one occurring.

- The *multiplicative law for independent events* says that   P(A and B) = P(A)P(B).

- The terms "mutually exclusive" and "independent" are sometimes confused. If events A and B are mutually exclusive then P(A and B) = 0. If events A and B are independent then P(A and B) = P(A)P(B).

- In general, to calculating the probability of an event
    * Define the experiment and list the outcomes in the sample space.
    * Assign probabilities to the outcomes so that each is between 0 and 1 and they add to 1.
    * List the outcomes of the event of concern.
    * Sum the probabilities of the outcomes that are in the event of concern.

- A random variable is a rule that assigns a numerical value to the outcomes of an experiment. If the sample space of values of the random variable is a discrete set of values, then the random variable is called discrete. If the sample space of values is a continuous set of values, then it is called a continuous random variable. You must understand the concept of a random variable and be able to distinguish between discrete and continuous random variables.

- The *probability distribution* of a discrete random variable is a table or function that lists the values of the variable and the probability with which it assumes those values. This table can be used to find probabilities of events. You must be prepared to give the probability distribution of discrete random variables. In particular you should know the characteristics of the binomial experiment and be able to find probabilities associated with the binomial distribution.

- A *binomial experiment* is an experiment consisting of several independent trials in which each trial results in either a success or a failure. The *binomial random variable* is a discrete variable that represents the number of successes in a binomial experiment. Probabilities associated with the binomial random variable can be found in the binomial probability tables in the appendix.

- The *mean* of a random variable is an average value of the random variable. It is the value that would balance the distribution. The *standard deviation* is a measure of the variability associated with the random variable. An interpretation of the amount of variability of a random variable can be given with either the *empirical rule* or *Chebyshev's rule*. The empirical rule applies to bell shaped distributions and Chebyshev's rule applies to any distribution.

- The *normal random variable* is a continuous random variable that is associated with the measurements of some numerical variable. The normal probability density curve (that corresponds to probabilities) is a bell shaped curve that is symmetrical. The mean and standard deviation are the parameters that distinguish between normal random variables. You should be familiar with the characteristics of the *normal distribution* and be able to find probabilities associated with the normal distribution.

- Probabilities associated with the normal distribution can be found by determining how far a score is from the mean. To do this, calculate the z-score and then finding the associated probability in the standard normal tables (z-tables) in the appendix.

- In general, to work probability problems for normally distributed populations
  * Draw a graph of a normal curve, label the mean, and shade the desired area.
  * Find the number of standard deviations a score is from the mean by finding the z-score.
  * Find the associated probability for the z-score in the standard normal probability table.
  * Relate the result to the problem at hand.

- There are several common distribution shapes that appear in the daily practice of statistics. Being able to detect the shape of the distribution from sample information is a very important task. For example, detecting normality is of utmost importance. The *midsummaries* tell us about symmetry, the *boxplot* indicates the length of the tails of a distribution, and the *normal probability plot* gives us a graphical means for checking normality.

- The normal probability plot is a graph of the sample data against the values we would have expected had the sample come from a normal population. If the sample is indeed from a normal population, then the graph should be close to a straight line. Deviations from a straight line indicate non-normality. You should be familiar with the patterns of normal probability plots for detecting skewness, short-tailed, and long tailed distributions.

Glossary

An **experiment** is the process of making an observation or taking a measurement.

The collection of all possible outcomes of an experiment is called the **sample space**, **S**.

Any subset of the sample space is called an **event**. An event is said to **have occurred** if any one of its elements is the outcome when the experiment is conducted.

The **probability of an event A** is the sum of the probabilities of the outcomes in A. We write it as P(A).

Events A and B are said to be **mutually exclusive** if they have no outcomes in common.

The Additive Law for Mutually Exclusive Events states

Let A and B be two mutually exclusive events, then   P(A or B) = P(A) + P(B)

The **complement of event A** is the collection of all outcomes in the sample space that are not in A.

The Complement Law states

Let A be an event with probability denoted by P(A), then $P(A^c) = 1 - P(A)$.

The Additive Law states

Let A and B be any two events, then $P(A \text{ or } B) = P(A) + P(B) - P(A \text{ and } B)$

Events A and B are said to be **independent** if the occurrence of one does not affect the probability of the occurrence of the other. Otherwise, A and B are **dependent**.

The Multiplication Law for Independent Events states

Let A and B be two independent events, then $P(A \text{ and } B) = P(A)P(B)$

The **conditional probability** of event A given that event B has occurred is found by dividing the probability that both A and B occur by the probability that B occurs. That is,

$$P(A|B) = \frac{P(A \text{ and } B)}{P(B)}$$

provided that P(B) is not zero.

The Multiplication Law states

Let A and B be two events with nonzero probability, then $P(A \text{ and } B) = P(A|B)P(B)$.

A **random variable**, which is denoted by a letter such as x or y, is a rule that represents the possible numerical values associated with the outcomes of an experiment. The list of values will be called the *sample space* or the *range* of the random variable.

A table or function that lists all of the possible values of a discrete random variable and their associated probabilities is called the **probability distribution** of the random variable.

Empirical Rule Applied To Population Distributions

If the probability distribution is bell-shaped, then

1. Approximately 68% of the distribution will fall within one standard deviation of the mean. That is, 68% of the distribution is between $\mu - \sigma$ and $\mu + \sigma$.

2. Approximately 95% of the distribution will fall within two standard deviations of the mean. That is, 95% of the distribution is between $\mu - 2\sigma$ and $\mu + 2\sigma$.

3. Essentially all of the distribution will fall within three standard deviations of the mean. That is, essentially all of the distribution is between $\mu - 3\sigma$ and $\mu + 3\sigma$.

Chebyshev's Rule

Regardless of the shape of the distribution we have

1. At least 3/4 of the distribution will fall within two standard deviations of the mean. That is, from $\mu - 2\sigma$ to $\mu + 2\sigma$.

2. At least 8/9 of the distribution will fall within three standard deviations of the mean. That is, from $\mu - 3\sigma$ to $\mu + 3\sigma$.

A **Bernoulli population** is a population in which each element is one of two possibilities. The two possibilities are usually designated as either success or failure. A **Bernoulli trial** is observing one element in a Bernoulli population.

A **binomial experiment** is an experiment that consists of n repeated independent Bernoulli trials in which the probability of success on each trial is π and the probability of failure on each trial is $1-\pi$.

The random variable x, which gives the number of successes in the n trials of a binomial experiment, is called a **binomial random variable**. The sample space of values of x will be $S_x = \{0, 1, 2, \text{---}, n\}$

A **probability density function**, f(y), which describes the probability distribution for a continuous random variable y, has the following properties:

a) $f(y) \geq 0$
b) The total area under the probability density curve is 1.00, which corresponds to 100%
c) $P(a \leq y \leq b)$ = area under the probability density curve between a and b.

One of the most commonly observed random variables is one whose probability density curve is characterized by its mean and variance; it is called the **normal random variable**. Its probability density curve is symmetrical and bell-shaped.

The **z-score:** $z = (\text{score} - \mu)/\sigma$ gives the number of standard deviations that a score is from the mean. A negative z-score indicates that the score in question is below the mean and a positive z-score indicates that the score is above the mean. The distribution of z has a mean of 0 and a standard deviation of 1 and is known as the *standard* normal distribution.

The **pth percentile** is the value in the population, so that p% of the distribution lies below it.

A **normal probability plot** is a graph of the sample data against the values we would have expected had the sample come from a normal population.

Practice Test 1

I. Multiple Choice

1. Suppose an experiment consist of rolling a pair of dice. The number of elements in the sample space is

      a) 6   b) 8   c) 12   d) 36

2. Suppose events A and B are such that $P(A) = .3$ and $P(B) = .6$ and they are mutually exclusive. What is the probability of A or B?

      a) .9   b) .72   c) .18   d) none of the above.

3. In a room there are 3 men and 7 women. Two people are selected to leave the room. What is the probability both are women?

      a) 49/100   b) 7/10   c) 2/7   d) 7/15

4. If two events cannot happen at the same time then
a) they are independent.
b) the occurrence of one does not affect the chance of the other occurring.
c) they are mutually exclusive.
d) all of the above.

5. If the occurrence of one event does not affect the chance of another event occurring then
a) the two events are mutually exclusive.
b) both cannot happen at the same time.
c) the two events are independent.
d) all of the above.

6. Suppose a $2.00 slot machine pays $15.00 10% of the time. If you play 10 times
a) you will win at least $15.00.
b) you will win at least $5.00.
c) you will lose no more than $5.00.
d) you may lose $20.00.

7. Suppose the random variable y denotes the sum of the two up faces on a pair of dice that have been tossed.
a) The values of y are equally likely.
b) The probability distribution is symmetric.
c) The random variable is a binomial variable.
d) The random variable is continuous.

8. Suppose x is a binomial random variable with $n = 5$ and $\pi = .4$ then
a) the expected value of x is 2.
b) the variance of x is 1.2.
c) x gives the number of successes in 5 trials.
d) all of the above.

9. Suppose x is a binomial random variable with n = 5 and π = .4 then the probability that x is less than 3 is

a) .913   b) .683   c) .230   d) .346

10. Suppose x is a binomial random variable with n = 20 and π = .7 then the probability that x is greater than 16 is

a) .108   b) .238   c) .130   d) .072

11. Suppose x is a binomial random variable with n = 16 and π = .8 then the standard deviation of x is

a) 12.8   b) 2.56   c) 3.58   d) 1.6

12. Suppose x is a normal random variable with μ = 10 and σ = 2. The probability that x exceeds 15 is

a) .4938   b) .9938   c) .0062   d) 2.5

13. Suppose x is a normal random variable with μ = 50 and σ = 15. The probability that x is between 60 and 80 is

a) .2286   b) .7258   c) .2742   d) .2714

14. Suppose x has distribution given by

```
x    | 1   2   3   4   5
-----|-------------------
p(x) | .2  .4  .3  .1  .1
```

a) The probability that x is odd is .6.
b) The probability that x is greater than 3 is .2.
c) The probability that x is less than or equal to 2 is .6.
d) None of the above because the table is not a probability distribution.

15. Suppose x has distribution given by

```
x    | 1   2   3   4
-----|----------------
p(x) | .1  .3  .4  .2
```

The expected value of x is

a) 2.5   b) .35   c) 1.2   d) 2.7

16. Suppose x is the number of times a 1 comes up on the roll of a fair die 5 times.

a) x is a binomial random variable.
b) x is a normal random variable.
c) x is neither binomial nor normal.
d) x is continuous.

17. Scores on the Stanford-Binet IQ test are approximately normally distributed with mean 100 and standard deviation 15. So about 95% of the population have IQ's between

a) 85 and 115.
b) 70 and 130.
c) 55 and 145.
d) can't tell from the information given.

II. Fill in the blank

18. In order to apply the Empirical rule, the distribution should appear _____ _____.

19. A _____ population is a population in which each element is one of two possibilities.

20. The _____ of a random variable is the list of values it can assume.

21. A subset of the sample space is called an _____.

22. The _____ _____ is a list of all possible outcomes of an experiment.

III. Problems

23. Suppose 40% of the candidates for veterinarian school get their degree on time. If 3 candidates are randomly selected
a) List the elements in the sample space of outcomes.
b) Assign probabilities to the outcomes in the sample space.
c) What is the probability that two of the three graduate on time?

24. Suppose a sample space for an experiment is S = {a, b, c, d} where the outcomes {a} and {b} are equally likely and events {c} and {d} are equally likely but event {a} is twice as likely as event {c}.
a) What is the probability of event {a, c}?
b) What is the probability of event {a, c, d}?

25. Suppose a fourth of all football players drafted by professional teams play at least two seasons. From a random sample of 4 drafted players, what is the probability that
a) none return for their second year?
b) all return for their second year?
c) How many are expected to return for their second year?

26. Suppose there is a 50% chance of rain on day A and a 40% chance of rain on day B. Assume that the possibility of rain on the two days are independent.
a) What is the probability it rains both days?
b) What is the probability it rains neither day?

27. Suppose x is a random variable with probability distribution

x	0	2	4	6
p(x)	1/7	2/7	3/7	1/7

72

a) What is the probability that x is odd?
b) What is the probability that x less than 3?
c) What value of x is most likely to happen?
d) What is E(x)?

28. Suppose 20% of the drinkers of soft drinks prefer uncola to cola. Let x be the number of drinkers of uncola out of 4 randomly selected drinkers of soft drinks. Construct the probability distribution table for the random variable x.

29. The life of a dishwasher is assumed to be normally distributed with a mean of 12 years and a standard deviation of 18 months. What percent will last
a) more than 15 years?
b) between 8 and 11 years?
c) Ninety percent of all dishwashers will expire within how many years?

30. Suppose the mean GPA at the end of the freshman year is 2.5 with a standard deviation of 0.8. Assuming the scores are normally distributed,
a) what percent have GPAs between 2.0 and 3.5?
b) what percent have GPAs above 3.4?

31. Suppose X is a discrete random variable with probability distribution table as follows:

```
x    |  0   1   2   3   4
-----|---------------------
p(x) | .1  .2  .4  .2  .1
```

a) What value of X is most likely to come up?
b) What is the probability X will be odd?
c) Graph the distribution.

32. Classify each of the following random variables as either continuous or discrete.
a) The number of defective computers in a shipment of 100 computers.
b) The useful lifetime of a video cassette recorder.
c) The reading proficiency test score for 4th graders.
d) The number of left-handed people in a class of 40 students.

33. Suppose 60% of all adult males are overweight. What is the probability that out of a random sample of 15 male adults, more than 10 are overweight?

34. Suppose the mean reading score of 4th graders on a standardized reading test is 65 and the standard deviation is 15. Assuming the scores are normally distributed,
a) Calculate the probability that a randomly selected 4th grader would score at least 70.
b) Calculate the probability that a randomly selected 4th grader would score between 40 and 75.

35. Suppose 40% of the members of a social club contribute regularly to a local charity. From a random sample of 8 members of the social club, what is the probability that more than half would contribute to the charity?

36. The daily weight gain of steers in a western feedlot is assumed to be normally distributed with a mean of 2.2 pounds and a standard deviation of .8 pounds. What percent of the steers will gain
a) more than 4 pounds per day?
b) between 1 and 3 pounds per day?
c) Ninety percent of all steers will gain at least how many pounds per day?

Answers to Practice Test 1

1. d 2. a 3. d 4. c 5. c 6. d 7. b 8. d 9. b 10. b 11. d 12. c 13. a 14. d 15. d 16. a 17. b 18. bell-shaped 19. bernoulli 20. range 21. event   22. sample space

23. a) {(sss), (ssf), (sfs), (fss), (sff), (fsf), (ffs), (fff)}
b) .064 .096 .096 .096 .144 .144 .144 .216
c) 3(.096) = .288

24. a) 1/2 b) 2/3

25. a) $(3/4)^4$ b) $(1/4)^4$ c) 1

26. a) .2 b) .3

27. a) 0   b) 3/7   c) 4   d) 22/7

28.

probability distribution

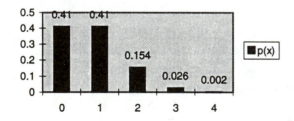

29. a) .0228   b) .2476   c) 13.92

30. a) .6301   b) .1292

31. a) 2   b) .4

c)

probability distribution

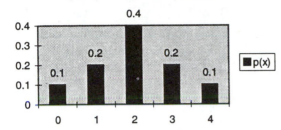

32. a) discrete   b) continuous   c) continuous   d) discrete

33. .217   34. a) .3707   b) .1011   35. .174

36. a) .0122   b) .7745   c) 1.176 pounds

Practice Test 2

I. Multiple choice

1. Suppose an experiment consist of rolling a die and tossing a coin. The number of elements in the sample space is
        a) 6   b) 8   c) 12   d) 36

2. Suppose events A and B are such that $P(A) = .5$ and $P(B) = .4$ and they are independent. What is the probability of A or B?
        a) .9   b) .2   c) .7   d) none of the above.

3. A bowl contains 4 red balls and 6 white balls. Two balls are drawn without replacement. What is the probability both are white?
        a) 6/10   b) 4/10   c) 1/3   d) 36/100

4. If two events are independent then
a) both cannot happen at the same time.
b) the occurrence of one does not affect the chance of the other occurring.
c) they are mutually exclusive.
d) all of the above.

5. If two events are mutually exclusive then
a) both cannot happen at the same time.
b) the occurrence of one does not affect the chance of the other occurring.
c) they are independent.
d) all of the above.

6. If there are 10 multiple choice questions on a test, each with four choices and you guess on each one, what is the probability you get all right?

       a) .10   b) $.4^{10}$ c) $.10^4$ d) $.25^{10}$

7. Suppose the probability of an event is .95. Then the event
a) will occur 95 times out of every 100 times the experiment is conducted.
b) may never occur.
c) will always occur.
d) is certain to occur in the first ten times the experiment is conducted.

8. Suppose an experiment consist of tossing 3 coins. The number of elements in the sample space is
a) 3   b) 8   c) 9   d) 27

9. The probabilities assigned to the outcomes in an experiment must
a) be greater than 0.
b) be less than 1.
c) add to 1.
d) all of the above.

10. Suppose a box has 4 red and 6 white balls in it. Three balls are selected with replacement and the colors recorded. Let the random variable y denotes the number of white balls drawn.
a) The values of y are equally likely.
b) The probability distribution is symmetric.
c) The random variable is a binomial variable.
d) The random variable is continuous.

11. Suppose x is a binomial random variable with n = 8 and π = .5 then
a) the expected value of x is 4.
b) the variance of x is 2.
c) x gives the number of successes in 8 trials.
d) all of the above.

12. Suppose x is a binomial random variable with n = 8 and π = .5 then the probability that x is less than 3 is
a) .219 b) .144 c) .363   d) .856

13. Suppose x is a binomial random variable with n = 15 and π = .2 then the probability that x is greater than 4 is
a) .164 b) .836 c) .188   d) .103

14. Suppose x is a normal random variable with μ = 50 and σ = 10. The probability that x exceeds 65 is
a) .4332   b) .9332   c) .0668   d) .5668

15. Suppose x has distribution given by

```
x    |  1   3   5   7   9
---- |---------------------
p(x) | .1  .2  .4  .3  .1
```

a) The probability that x is odd is .1.
b) The probability that x is greater than 5 is .4.
c) The probability that x is less than or equal to 4 is .3.
d) None of the above because the table is not a probability distribution.

16. Scores on a standardized test are approximately normally distributed with mean 500 and standard deviation 100. So about 95% of the distribution lies between
a) 400 and 600.
b) 300 and 700.
c) 200 and 800.
d) can't tell from the information given.

II. Fill in the blank

17. A _____ experiment consists of n repeated independent trials, each of which results in one of two outcomes.

18. A _____ _____ assigns numerical values to the outcomes of an experiment.

19. A table that lists the possible values of a discrete random variable and the associated probabilities is called a _____ _____ table.

20. A _____ random variable assumes all values in a line interval.

21. The Empirical rule was derived from the _____ distribution.

III. Problems

22. Classify each of the following random variables as either continuous or discrete.
a) The time to dismantel a rifle.
b) The number of times a record is played by a disc jockey on an evening radio program.
c) The number of points scored by a football team during the season.
d) The duration of a piano concert.

23. Suppose x is a discrete random variable with probability distribution table as follows:

```
x    |  2   5   7   8
---- |----------------
p(x) | .2  .4  .3  .1
```

77

a) What value of x is most likely to come up?
b) What is the probability x will be odd?
c) Graph the distribution.

24. Suppose 30% of all drivers speed when traveling on an interstate highway. What is the probability that out of 15 randomly selected drivers fewer than 5 speed on interstates?

25. Suppose a distribution which is somewhat bell-shaped has a mean of 25 and a standard deviation of 10.
a) What are the limits that would contain approximately the middle 95% of the distribution?
b) Would it be unusual to see a score above 45? Why or why not.

26. At each of three doors there are two keys one of which will open the door. If you pick the right key on all three doors you win a new car.
a) List the elements in the sample space of outcomes.
b) Assign probabilities to the outcomes in the sample space.
c) What is the probability you win the car?

27. Suppose a sample space for an experiment is S = {a, b, c, d} where the outcomes are equally likely.
a) What is the probability of event {a, b}?
b) What is the probability of event {a, c, d}?

28. Suppose 60% of the student body attends the football game. If 3 students are randomly selected, what is the probability that
a) all attend the game?
b) at least 2 attend the game?
c) none attend the game?

29. A third of all former convicts return to jail within a year. From a random sample of 4 former convicts, what is the probability that
a) none return within a year?
b) exactly 2 return within a year?
c) all return in a year?
d) What is the expected number to return in a year?

30. If one shoplifts they stand a 25% chance of being caught. Suppose there are 4 shoplifters in a store. What is the probability that
a) none are caught?
b) at least 2 are caught?
c) all are caught?
d) What is the expected number to be caught?

31. Forty percent of all university students live in dorms. For a survey, four students are selected randomly. What is the probability that
a) none live in dorms?
b) at least 2 live in dorms?

c) all live in dorms?
d) Out of the four, how many would you expect to live in dorms?

32. In the NBA slam-dunk contest suppose a participant makes 3 attempts at the 360 degree rotation slam in which he has a 30 percent chance of making.
a) List the elements in the sample space of outcomes.
b) Assign probabilities to the outcomes in the sample space.
c) What is the probability that he makes at least 2 of the 3 attempts?

33. Suppose events A and B are independent and P(A) = .4 and P(B) = .5. Find P(A or B).

34. A fair die is rolled. If the outcome is less than 3 a coin is tossed, otherwise the die is tossed again. Are the outcomes equally likely? Why?

35. Three employees of a company are late to work 20% of the time. Suppose they independently choose to be either late (L) or prompt (P).
a) List the sample space of possibilities for any randomly selected day.
b) Assign probabilities to the outcomes.
c) What is the probability that no more than one is late to work?

36. Suppose there is a 30% chance of a pop quiz in class A and a 20% chance in class B.
a) What is the probability of a pop quiz in both classes?
b) What is the probability of a pop quiz in either class A or class B?

Answers to Practice Test 2

1. c 2. c 3. c 4. b 5. a 6. d 7. b 8. b 9. d 10. c 11. d 12. b 13. a 14. c 15. d 16. b 17. binomial 18. random variable 19. probability distribution 20. continuous 21. normal
22. a) continuous b) discrete c) discrete d) continuous
23. a) 5 b) .7
c)

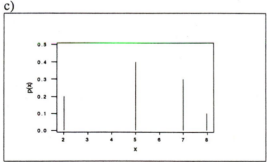

24. .517   25. a) (5,45) b) yes, the chance is only about .025
26. a) {(sss), (ssf), (sfs), (fss), (sff), (fsf), (ffs), (fff)}   b) 1/8 each   c) 1/8

27. a) .5   b) .75

79

28. a) .216   b) .648   c) .064

29. a) .198   b) .296   c) .012   d) 1.333

30. a) .3164   b) .262   c) .004   d) 1

31. a) .130   b) .526   c) .026   d) 1.6

32. a) {(sss), (ssf), (sfs), (fss), (sff), (fsf), (ffs), (fff)}
b) .027, .063, .063, .063, .147, .147, .147, .343   c) .216

33. .7

34. No, a (1,H) has probability 1/12 and (1,1) has probability 1/36.

35. a) {(PPP), (PPL), (PLP), (LPP), (PLL), (LPL), (LLP), (LLL)}
b) .512, .128, .128, .128, .032, .032, .032, .008   c) .896

36. a) .06   b) .44

5

Sampling Distributions

Statistics involves the study of the important population characteristics. Usually this means that a sample is selected and used to estimate specific population parameters. A main task then, is deciding which statistic from the sample best estimates the unknown parameter. To help in this decision we must understand the concept of the sampling distribution of a statistic.

Important issues—Chapter 5

* The value assumed by a statistic is dependent on the observed sample. Because a large number of possible samples can be chosen from a population, there is a large number of different values that a statistic can realize. The distribution of the potential values that the statistic can assume from the various samples is called the *sampling distribution* of the statistic. One of your main tasks is to understand the basic concepts of a sampling distribution.

* Important statistics whose sampling distribution is studied in this chapter are

  * $\bar{y}$ - the sample mean
  * $\bar{y}_T$ - a trimmed sample mean
  * M - the sample median
  * s - the sample standard deviation
  * p- the sample proportion
  * r - the sample correlation coefficient
  * $\bar{y}_1 - \bar{y}_2$ - the difference of two sample means
  * $M_1 - M_2$ - the difference of two sample medians
  * $p_1 - p_2$ - the difference of two sample proportions

* Take, for example, the sample mean $\bar{y}$. Does it provide a good estimate of the population mean μ? If its potential values are not too erratic and are somewhat *stable* from sample to sample then the answer is yes. If, however, the statistic assumes widely varying values from sample to sample, then the conclusions we draw about the population parameter will be unreliable. It is through the variability of the sampling distribution of the statistic, as measured by its *standard deviation,* that we determine how precise an estimate of a parameter will be. Understanding how the standard deviation of a statistic affects the different possible values of the statistic is of primary importance to you.

- As with any distribution, describing the sampling distribution of a statistic involves three things.

    * describe the general shape of the distribution.
    * give a measure of where the distribution is centered.
    * give a measure of the amount of variability in the distribution.

- One of the most important theorems in statistics, called the *Central Limit Theorem* (CLT), resolves these issues concerning the sampling distribution of the sample mean $\overline{y}$. It says that for a sufficiently large sample size, the sampling distribution of $\overline{y}$ is approximately normally distributed with a mean of μ and a standard deviation of $\sigma/\sqrt{n}$. Having completed this chapter, you should be able to state and understand the Central Limit Theorem.

- The Central Limit Theorem also applies to the sampling distribution of the sample proportion, p. It says that for a sufficiently large sample size, the sampling distribution of p is approximately normally distributed with a mean of π and a standard deviation of $\sqrt{\pi(1-\pi)/n}$.

- Knowledge of the sampling distribution of a statistic allows us to determine which of the potential values of the statistic are reasonable and which are unlikely. Thus when a value is obtained from a particular sample, knowing the sampling distribution of the statistic allows us to determine immediately if the value is in agreement with our ideas about the population. Also, probability problems involving the statistic can be worked if we know its sampling distribution.

- Simulations are used in this chapter to better understand the properties of the sampling distribution of a statistic. For example, simulations are used to compare the sampling distribution of $\overline{y}$ to the sampling distributions of the sample median and the sample midrange. Any of the three could conceivably be used to estimate the center of a distribution; using simulations under different population conditions, we are able to determine the best statistic to use in a variety of situations. You should be able to describe the differences between the sampling distributions of the mean, median, and midrange.

- In the study of statistics you will be asked to compare the difference between two parameters. To accomplish this you must study the sampling distribution of the difference between two statistics. For example, to compare two population means we need to compare the sample means that are obtained from two independent random samples taken from the two respective populations. To assess how close the difference in the two sample means will be to the difference in the two population means, we need to investigate the sampling distribution of $\overline{y}_1 - \overline{y}_2$ where $\overline{y}_1$ is the mean of a random sample from population 1 and $\overline{y}_2$ is the mean of an independent random sample from population 2. The same applies to the difference between two sample proportions.

Glossary

The **sampling distribution of a sample statistic** is the probability distribution associated with the various values that the statistic can assume in repeated sampling.

The **sampling distribution of** $\bar{y}$ **when sampling from a normally distributed population** that has mean μ and standard deviation σ is exactly normally distributed, centered at μ and has a standard deviation of $\sigma/\sqrt{n}$.

When sampling from general populations, the **Central Limit Theorem** says if the sample size n is sufficiently large, then the sampling distribution of $\bar{y}$ is **approximately** normally distributed. The approximation becomes better as the sample size increases.

The Central Limit Theorem applied to p says if the sample size n is sufficiently large, then the sampling distribution of p will be approximately normally distributed. Furthermore, its distribution is centered at π, the true proportion of successes in the Bernoulli population, and has a standard deviation of $\sqrt{\pi(1-\pi)/n}$.

Normal approximation of the Binomial Distribution

When n becomes large, the binomial distribution can be reasonably approximated with a normal distribution with a mean of $n\pi$ and a standard deviation of $\sqrt{n\pi(1-\pi)}$. For a good approximation n should be large enough that both $n\pi > 5$ and $n(1-\pi) > 5$.

Sampling Distribution of $\bar{y}_1 - \bar{y}_2$

Two independent random samples of size n_1 and n_2 are selected from two populations that have means μ_1 and μ_2 and standard deviations σ_1 and σ_2, respectively. If n_1 and n_2 are each larger than 30, the sampling distribution of the difference in the sample means, $\bar{y}_1 - \bar{y}_2$, is approximately normally distributed, centered at $\mu_1 - \mu_2$, and has a standard deviation of $\sqrt{\sigma_1^2/n_1 + \sigma_2^2/n_2}$

Sampling Distribution of $p_1 - p_2$

If n_1 and n_2 are sufficiently large, then the sampling distribution of $p_1 - p_2$ is approximately normally distributed, centered at $\pi_1 - \pi_2$, and has a standard deviation of $\sqrt{(\pi_1(1-\pi_1)/n_1) + (\pi_1(1-\pi_2)/n_2)}$

Practice Test 1

I. Multiple Choice

1. The Central Limit Theorem describes
a) the distribution of the parent population.
b) the distribution of all possible samples.

c) the distribution of all possible values of the sample mean.

d) none of the above.

2. The standard deviation of a statistic describes

a) the shape of its distribution.

b) the center of its distribution.

c) the amount of skewness associated with its distribution.

d) the amount of variability associated with its distribution.

3. Knowing the sampling distribution of a statistic allows one

a) to evaluate a value of the statistic once it is calculated from a sample.

b) to compute probabilities associated with the statistic.

c) to determine if the obtained value of a statistic is consistent with their views about the population.

d) all of the above.

4. As sample size increases the standard deviation of $\bar{y}$

a) becomes smaller.

b) becomes larger.

c) remains the same.

d) becomes unstable.

5. For a sample size of 400, the maximum standard deviation of p is

a) .00125

b) .000625

c) .025

d) .0125

6. The sampling distribution of $\bar{y}$ is normal or approximately normal when

a) the parent distribution is normal and the sample size is small.

b) the parent distribution is normal and the sample size is large.

c) the parent distribution is not normal but the sample size is large.

d) all of the above.

7. Suppose a random sample of size 100 is taken from a Bernoulli population with 30% successes. What is the probability of a sample proportion less than 25%?

a) .8621

b) .3621

c) .1379

d) .4583

8. Suppose a random sample of size 81 is taken from a population with mean 250 and a standard deviation of 50. What is the probability that the sample mean is less than 260?

a) .4641

b) .9641

c) .0359

d) 1.8

II. Problems

9. Identify the following statistics.
a) In 1980 the crime rate for Charlotte, NC was 17.6 per 100,000.
b) Last semester 32% of the freshman class had a GPA of 3.0 or better.
c) The average reaction time of 30 senior citizens taken on an instrument at the drivers license bureau is 84.63.
d) The middle salary of 25 new Ph.D. psychologists was $25,200.
e) The lowest aggressive tendency score by 28 teenagers was 11.

10. The mean life span of the people in a particular section of Siberia is 85 years with a standard deviation of 10 years. A random sample of 25 of these people is selected. Approximate
    the probability that their average age is
a) less than 90 years?
b) over 100 years?
c) Using the empirical rule we would feel reasonably sure the mean life span would fall between what two values?

11. It has been reported that at least 10% of the adult male population suffers from kidney stones. A random sample of 900 males is selected and tested for kidney stones. If p denotes the sample
    proportion of adult males who suffer from kidney stones, describe its sampling distribution.

12. In Exercise 11 the random sample showed that 81 had symptoms of kidney stones. Does this evidence cast doubt on the claim that at least 10% of the adult male population suffers from kidney stones?

13. Suppose the mean GPA at the end of the freshman year is 2.6 with a standard deviation of 0.6. Is it unusual for a random sample of 144 students to end their freshman year with an average of 2.75? Why? Calculate the probability that their average would be at least 2.75

14. Suppose 40% of all entering freshmen never get a bachelors degree. What is the probability that out of a random sample of 400 students fewer than 180 never get a bachelors degree?

15. In Exercise 14, how large of a sample is necessary so that the standard deviation is no more than .02?

16. In a previous exercise (Exercise 36, Practice Test 1, Chapter 4) it was assumed that the daily weight gain of steers in a western feedlot had an average of 2.2 pounds and a standard deviation of .8 pounds. Suppose a rancher has 36 steers in the feedlot.
a) What is the probability that his steers will have an average daily weight gain greater than 2.5 pounds?
b) What is the probability that they have an average daily weight gain between 1.5 and 2.0 pounds?

Answers to Practice Test 1

1. c 2. d 3. d 4. a 5. c 6. d 7. c 8. b
9. a) sample rate b) sample proportion c) sample mean d) sample median e) sample low score
10. a) .9938   b) 0⁻   c) 81 - 89
11. The sampling distribution of p is approximately normally distributed with mean .10 and standard deviation .01.
12. No, z = -1 which is not unusual
13. Yes, z = 3, .0013   14. z = 2.04, .0207   15. 625
16. a) .0122   b) .0668

Practice Test 2

I. Multiple Choice

1. The variability in a sampling distribution is described by
a) its location measure
b) its standard deviation
c) its long tails
d) its skewness

2. The general shape of the distribution of a statistic is described by
a)   the parent distribution
b)   the variability of the parent distribution
c)   the distribution of the sample
d)   the sampling distribution of the statistic

3. For the Central Limit Theorem to properly described the sampling distribution of $\overline{y}$
a)   the parent distribution must be normal
b)   the sample size must be sufficiently large
c)   the standard deviation of the parent distribution must be given
d)   all of the above

4. If a sample of size 100 is taken from a population with mean 60 and a standard deviation of 10 then the average of the sample should be between
a)   30 and 90
b)   40 and 80
c)   50 and 70
d)   58 and 62

5. Suppose a random sample of size 400 is taken from a Bernoulli population with 80% successes. What is the probability of a sample proportion greater than 85%?
a)   .0062
b)   .4938
c)   .9938

d) .5062

6. Suppose a random sample of size 64 is taken from a population with mean 75 and a standard deviation of 24. What is the probability that the sample mean assumes a value less than 78?
a)   .3413
b)   .1587
c)   .8413
d)   .5587

7. Which of the following are statistics?
a) μ
b) $\overline{y}$
c) θ
d) π

8. To reduce the standard deviation of $\overline{y}$ by one-half the sample size will have to be
a) halved.
b) doubled.
c) tripled.
d) quadrupled.

II. Problems

9. An anxiety test has a mean of 25 and a standard deviation of 5.
a) Approximate the probability that a random sample of 100 taking the test will have an average greater than 26.2.
b) Using the empirical rule we would feel reasonably sure that their average would fall between what two values?

10. A farmer raises Thanksgiving turkeys that have an average weight of 18 pounds with a standard deviation of 4 pounds. He randomly chooses 64 turkeys for shipment.
a) What is the probability that they have an average weight greater than 19.4 pounds?
b) What is the probability that they have an average weight between 19 and 20 pounds?

11. Suppose a Bernoulli population has p proportion of successes. Let p denote the sample proportion obtained from a sample of size n.
a) What is the maximum standard deviation of p when the sample size is 500?
b) What sample size is necessary so that the maximum standard deviation of p is no more than .01?

12. Suppose the contraction rate for a certain children's disease (percent of those exposed that contract the disease) is 35%. What is the probability that out of a random sample of 50 exposed children fewer than 15 come down with the disease?

13. It has been reported that 45% of all households have microwave ovens. What is the probability that a random sample of 400 households would have more than half with microwave ovens?

14. The average amount of non-alcoholic beverage consumed by Americans is 162.85 gallons per person each year. Assume that the standard deviation is 40 gallons. What is the probability that the average amount consumed by a sample of 64 would be
a) less than 150 gallons?
b) between 155 and 165 gallons?

15. Suppose a population has a mean of 500 and a standard deviation of 100.
a) What is the standard deviation of the sample mean obtained from samples of size 100.
b) How large of a sample would be required to reduce the standard deviation to 5?

Answers to Practice Test 2

1. b 2. d 3. b 4. d 5. a 6. c 7. b 8. d
9. a) .0082   b) 24 and 26   10. a) .0026   b) .0228
11. a) .0224   b) 2500   12. .2296
13. z = 2.01, .0222   14. a) z = -2.57, .0051 b) .6082   15. a) 10 b) 400

6

Describing Distributions

Describing population distributions via collected data is the primary focus of this chapter. The diagnostic tools developed in previous chapters are used collectively to first determine the shape of the distribution and then help to decide which parameters best describe the characteristics of interest. Once the parameters are identified we next learn how to estimate them.

Important issues—Chapter 6

- To describe the parent distribution we are concerned with

  * the general *shape* of the distribution
  * a measure of the *center* of the distribution
  * a measure of *variability* of the distribution

- *Distributional shapes* are classified as near normal, skewed, symmetric short-tailed, symmetric long-tailed, and multimodal. A major task is identifying the various population distribution shapes.

- The *stem and leaf plot*, *histogram*, and *boxplot* give the general appearance of the collected data which, assuming we have a representative sample, resembles the shape of the parent distribution.

- The boxplot shows skewness, symmetry, and identifies outliers. The *midsummary analysis* reaffirms the skewness or the symmetry. The *normal probability plot* checks for normality as well as identifies other unusual behavior in the data.

- Once the shape of a distribution is determined we next choose a parameter that measures the population characteristic of interest. One important task, for example, is deciding whether it is better to examine the population mean or the population median.

- Once the parameter is decided upon, we are then in a position to determine what statistic best estimates the parameter. In making this decision we must be familiar with the desirable characteristics of an *estimator* (the statistic used to estimate the parameter).

- When the sampling distribution of an estimator is centered at the parameter being estimated we say that it is an *unbiased estimator*. Important unbiased estimators are $\overline{y}$ as an estimate of μ and s^2 as an estimate of σ^2.

- Restricting ourselves to unbiased estimators, we can say that the smaller the standard deviation, the more tightly clustered will be the potential values of the estimator about the unknown parameter. Thus it is important that an estimator have a small standard deviation.

- For symmetric distributions, the mean, the median, and all the trimmed means coincide; therefore to estimate a typical score we simply estimate the mean. The estimator used to estimate the mean is determined by the type of symmetric distribution we have observed.

- For skewed distributions, the population median is usually the measure of center used to represent a typical score. The sample median is used to estimate the population median. The distance between the first and third quartiles, the q-spread $= \theta_3 - \theta_1$, is the preferred measure of dispersion in skewed populations.

- A robust estimator is one that works well in a wide variety of population distributions. One such robust estimator of the center of a distribution is the trimmed mean, $\overline{y}_T$. It is the preferred estimator of the center of a long-tailed distribution because it is not adversely effected by outliers that characteristic of a long-tailed distribution.

- If the parent distribution is normal, there is no better estimator of μ than the sample mean, $\overline{y}$.

Glossary

The distribution of the measurements in the original population is called the **underlying** or **parent distribution**.

A **point estimator** of a parameter is a *statistic* whose values should be close to the true value of the parameter. The actual numerical value that the point estimator assumes from the collected data (the sample) is called the **point estimate**.

If the sampling distribution of an estimator has a mean equal to the parameter being estimated, then it is called an **unbiased estimator** of the parameter. If the mean of the sampling distribution is not equal to the parameter, the estimator is said to be **biased**.

Statistical procedures that are insensitive to departures from assumptions are called **robust**.

Estimating Parameters

- The sample mean, $\overline{y}$, is the recommended estimator of the population mean when the parent distribution is normal, near normal, or symmetric with tails that are *not* excessively long.

- A trimmed sample mean, $\overline{y}_T$, is the recommended estimator of the population mean when the parent distribution is symmetric with long tails.

- The sample median, M, is the recommended estimator of the population center when the parent distribution is skewed in either direction.

- The sample variance, s^2, is an unbiased estimator of the population variance, σ^2.

Practice Test 1

I. Multiple Choice

1. In tabulating the family incomes of subjects you would expect to find that
a) the mean income is greater than the median income, because the distribution of incomes is generally skewed right.
b) the mean income is less than the median income, because the distribution of incomes is generally skewed right.
c) the mean income is greater than the median income, because the distribution of incomes is generally skewed left.
d) the mean income is less than the median income, because the distribution of incomes is generally skewed left.

2. Suppose the midsummaries from a letter value display get progressively larger as you go down. This would suggest that the parent distribution
a) is skewed right.
b) is skewed left.
c) is normally distributed.
d) has long tails.

3. If the midsummaries from a letter value display show no distinct pattern then
a) the parent distribution is bimodal.
b) the parent distribution is skewed.
c) nothing conclusive can be said about the parent distribution.
d) the parent distribution is definitely not bimodal.

4. If a normal probability plot of the sample shows a straight line pattern then it is suggested that the parent distribution
a) has short tails.
b) is skewed in one direction.
c) has long tails.
d) is normally distributed.

5. If a normal probability plot of the sample shows an S shape then it is suggested that the parent distribution
a) has short tails.
b) is skewed in one direction.
c) has long tails.
d) is normally distributed.

6. The box plot can reveal
a) skewness.
b) long tails.
c) short tails.
d) all of the above.

7. The mean and median will be the same when
a) there are outliers on the left but not the right tail of the distribution.
b) the distribution has outliers evenly distributed on both tails.
c) there are no outliers.
d) the distribution is symmetric.

II. Problems

8. Identify the following parameters.
a) The highest score ever obtained on certain video game is 2,465,389.
b) The middle salary for a large group of employees in a certain corporation is $24,285.
c) The percent of American families that own more than two automobiles is 15%.
d) A hall of fame baseball player had a lifetime batting average of .324.
e) Only 20% of all new TV shows last more than 2 seasons.

9. A statistic is called an _____ estimator if its sampling distribution has a mean equal to the parameter being estimated.

10. Suppose you have a sample and wish to investigate a typical score in the parent population. What parameter would you study if a box plot of the sample suggested that the parent population is
a) approximately normal in shape?
b) has tails significantly shorter than the normal distribution?
c) has tails significantly longer than the normal distribution?
d) is highly skewed left?
e) is highly skewed right?

11. For the following two data sets construct stem and leaf plots and box plots to compare the distributions. Describe any noticeable differences between the two distributions.

Data set A: 81 73 86 90 75 80 75 81 85 87 83 75 70 65 80 76 64 74 86 80 83 67 82 78 76 83 71 90 77 81 82

Data set B: 87 77 66 75 78 82 82 71 79 73 91 97 89 92 75 89 75 95 84 75 82 74 77 87 69 96 65

12. Consider the data set that represents scores on a standardized algebra test for beginning calculus students.

    78 86 83 81 71 72 78 83 72 82 81 81 84 68 79 89 84 72 78 76 52
a) Construct a stem and leaf plot.
b) Construct a 5-number summary display.
c) Calculate all midsummaries.
d) Comment on the shape of the parent distribution.

13. A researcher is interested in studying the length of marriages that end in divorce. Following are the number of years of marriage obtained from a random sample of divorce records.

1.7, 2.6, 1.9, 6.5, 2.8, 3.1, 7.2, 15.8, 2.5, 3.4, 1.3, 0.2, 4.5, 2.4, 35.0, 7.5, 3.2, 2.1, 1.8, 5.0, 6.6, 4.2, 3.1, 12.8, 3.5, 2.9
a) From a stem and leaf plot comment on the distributional shape.
b) Based on the midsummaries how would you classify the distributional shape.
c) Construct a boxplot and comment on the distributional shape.
d) What measure do you think best represents the center of the distribution? What is the value it realized from the data?

14. A researcher for the national weather bureau is studying the intensity of thunderstorms in Kansas. Following are the intensity ratings of 25 thunderstorms in the summer months.

1.6, 2.4, 1.7, 0.7, 2.3, 3.1, 0.5, 1.6, 2.4, 3.6, 1.7, 4.2, 4.5, 2.6, 1.5, 0.8, 3.2, 2.1, 1.8, 3.0, 0.6, 4.2, 1.4, 1.2, 2.9
a) Construct a stem and leaf plot and letter value display including the midsummaries and the spreads.
b) From the midsummaries comment on the distributional shape.
c) Construct a boxplot and comment on the distributional shape.
d) What measure do you think best represents the center of the distribution? What is the value it realized from the data?

Answers to Practice Test 1

1. a 2. a 3. c 4. d 5. c 6. d 7. d
8. a) population high b) population median c) population proportion
   d) population mean   e) population proportion

9.   unbiased

10. a) population mean   b) population mean   c) trimmed population mean
        d) population median   e) population median

11. a)

```
Stem-and-leaf of A    N = 31          Stem-and-leaf of B      N = 27
Leaf Unit = 1.0                       Leaf Unit = 1.0

    2      6 45                           3      6 569
    3      6 7                            6      7 134
    3      6                             (8)     7 55557789
    5      7 01                          13      8 2224
    6      7 3                            9      8 7799
   10      7 4555                         5      9 12
   13      7 667                          3      9 567
   14      7 8
   (6)     8 000111
   11      8 22333
    6      8 5
    5      8 667
    2      8
    2      9 00
```

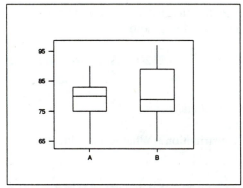

There is less variability in data set A. The median response in data set A is larger than the median
response in data set B. Data set A is possibly multimodal.

12.   a.

```
        5 | 2
        6 | 8
        7 | 1 2 2 2 6 8 8 8 9
        8 | 1 1 1 2 3 3 4 4 6 9
```

```
           b.                    c.
                      21         midsummaries    spreads
                 ---------------
        M(11) |        79     |     79
        Q(6)  |72          83 |     77.5          11
        E(3.5)|71.5        84 |     77.75         12.5
           R  |52          89 |     70.5          37
```

   d.

```
                 0      x-----------|_____|___|-----x
                                    |_____|___|

              -|----|----|----|----|----|----|----|----|-
              50   55   60   65   70   75   80   85   90
```

e. From the stem and leaf plot, midsummaries and the box plot it appears that the distribution is
skewed left.

13. a)
```
   Stem-and-leaf of C1       N = 26
   Leaf Unit = 1.0

   (18)    0  011112222223333344
      8    0  56677
      3    1  2
      2    1  5
      1    2
      1    2
      1    3
      1    3  5
```
The distribution is heavily skewed right.

b) **Letter Value Display**

	DEPTH	LOWER	UPPER	MID	SPREAD
N=	26				
M	13.5		3.150	3.150	
H	7.0	2.400	6.500	4.450	4.100
E	4.0	1.800	7.500	4.650	5.700
D	2.5	1.500	14.300	7.900	12.800
C	1.5	0.750	25.400	13.075	24.650
	1	0.200	35.000	17.600	34.800

The midsummaries indicate that the distribution is skewed right.

c) **Character Boxplot**

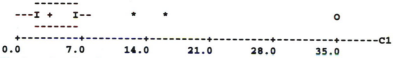

The boxplot also shows heavy skewness to the right.

d) Because of the heavy skewness, the median best represents the center of the distribution. Its value is M= 3.15

14. a)
```
   Stem-and-leaf of Intense   N = 25
   Leaf Unit = 0.10

      4     0  5678
      6     1  24
     12     1  566778
     (4)    2  1344
      9     2  69
      7     3  012
      4     3  6
      3     4  22
      1     4  5
```

Letter Value Display

	DEPTH	LOWER	UPPER	MID	SPREAD
N=	25				
M	13.0		2.100	2.100	
H	7.0	1.500	3.000	2.250	1.500
E	4.0	0.800	3.600	2.200	2.800
D	2.5	0.650	4.200	2.425	3.550
C	1.5	0.550	4.350	2.450	3.800
	1	0.500	4.500	2.500	4.000

b) The distribution is reasonably symmetric and possibly short tailed.

```
                         --------------------
           ------------I        +          I------------------
                         --------------------
       ------+---------+---------+---------+---------+---------+Intense
           0.80      1.60      2.40      3.20      4.00      4.80
```

d) Because the distribution is somewhat symmetric without long tails the best representation of the center of the distribution is the sample mean. Its value is $\overline{y} = 2.224$.

Practice Test 2

I. Multiple Choice

1. Suppose the midsummaries from a letter value display get progressively smaller as you go down. This would suggest that the parent distribution
a) is skewed right.
b) is skewed left.
c) is normally distributed.
d) has long tails.

2. Suppose the midsummaries from a letter value display are all about the same. This would suggest that the parent distribution
a) has short tails.
b) is symmetrical.
c) is normally distributed.
d) has long tails.

3. A box plot with no outliers on one tail and several outliers on the opposite tail indicates
a) skewness.
b) long tails.
c) short tails.
d) all of the above.

4. A box plot with no outliers is a definite indication that the parent distribution is not
a) symmetric
b) long tailed
c) normal
d) all of the above.

5. If a normal probability plot of the sample shows a straight line pattern on one tail and curves up sharply on the other tail then it is suggested that the parent distribution
a) has short tails.
b) is skewed in one direction.
c) has long tails.
d) is normally distributed.

6. If a normal probability plot of the sample shows an inverted S shape then it is suggested that the parent distribution
a) has short tails.
b) is skewed in one direction.
c) has long tails.
d) is normally distributed.

7. Regarding parameters and statistics
a) if the value of the parameter is known then the value of the statistic is known.
b) if the value of the statistic is known then the value of the parameter is known.
c) if the value of the statistic is known then the value of the parameter can only be estimated.
d) none of the above.

II. Problems

8. What parameter would best represent the parent population if a box plot of the sample had no outliers on the right tail and numerous outliers on the left tail?

9. What parameter would best represent the parent population if a box plot of the sample had numerous outliers on both tails?

10. Estimator A has _____ variability than estimator B if its standard deviation is smaller for the same sample size.

11. A psychological test for measuring racial prejudice is given to a random sample of 25 high school students. From the recorded data, construct a stem and leaf plot and box plot. Analyze the midsummaries and comment on the shape of the parent distribution.

   59, 54, 41, 51, 37, 42, 42, 35, 44, 46, 44, 41, 58, 58, 38, 47, 33, 48, 48, 45, 37, 29, 52, 61, 48

12. Following are the lengths (in seconds) for the violent acts in a horror movie:

   2.7, 3.6, 7.9, 6.3, 2.9, 3.5, 7.5, 15.3, 6.5, 10.4, 3.3, 1.2,
   4.3, 2.9, 36.0, 7.9, 3.6, 4.1, 1.8, 5.7, 6.2, 8.2, 4.1, 12.4,

a) Construct a stem and leaf plot.
b) Construct a 5-number summary display and include the midsummaries and spreads.
c) Calculate the sample mean and standard deviation.

13. A researcher for a long-distance telephone company is interested in studying the length of a typical long-distance phone call. Following are the results (recorded in minutes) of a random sample of 35 calls.

12.8, 3.5, 2.9, 9.4, 8.7, 3.5, 4.8, 7.7. 5.9, 6.2, 2.8, 4.7,1.7, 2.6, 1.9, 6.5, 2.8, 3.1, 7.2, 15.8, 2.5, 3.4, 1.3, 0.2, 4.5, 2.4, 35.0, 7.5, 3.2, 2.1, 1.8, 5.0, 6.6, 4.2, 3.1,

a) Calculate the midsummaries and comment on the distributional shape.

b) Construct a boxplot and comment on the distributional shape.

c) What measure do you think best represents the center of the distribution? What is the value it realized from the data?

Answers to Practice Test 2

1. b 2. b 3. a 4. b 5. b 6 a 7. c

8. population median   9. A trimmed population mean

10. less

11.
```
        2 | 9
        3 | 3 5 7 7 8
        4 | 1 1 2 2 4 4 5 6 7 8 8 8
        5 | 1 2 4 8 8 9
        6 | 1
```

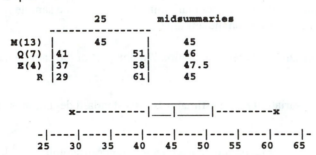

```
                     25              midsummaries
               ------------------
    M(13) |           45         |        45
     Q(7) |41               51|           46
     E(4) |37               58|           47.5
        R |29               61|           45


            x-----------|___|_____|---------x

            -|----|----|----|----|----|----|----|----|-
            25   30   35   40   45   50   55   60   65
```

The box plot and midsummaries suggest that the distribution does not deviate significantly from symmetry and may possibly even be normally distributed.

12. a)
```
        1 |2 8
        2 |7 9 9
        3 |6 5 3 6
        4 |3 1 1
        5 |7
        6 |3 5 2
        7 |9 5 9
        8 |2
        9 |            scores x 10⁻¹
       10|4
       11|
       12|4
       --|--------
       HI|153,360
```

b)
```
                     24            Midsummaries  Spreads
               ------------------
    M(12.5) |          5.0         |       5.0
     Q(6.5) |3.4              7.9|         5.65       4.5
     E(3.5) |2.8             11.4|         7.1        8.6
        R |1.2              36.0|         18.6        34.8
```

c) $\bar{y} = 7.0125$ $s = 7.05$

98

13. a) **Stem-and-leaf of minutes   N = 35**
    Leaf Unit = 1.0

```
  (22)    0  0111122222223333334444
   13     0  5566677789
    3     1  2
    2     1  5
    1     2
    1     2
    1     3
    1     3  5
```

Letter Value Display

	DEPTH	LOWER	UPPER	MID	SPREAD
N=	35				
M	18.0	3.500		3.500	
H	9.5	2.700	6.550	4.625	3.850
E	5.0	1.900	8.700	5.300	6.800
D	3.0	1.700	12.800	7.250	11.100
C	2.0	1.300	15.800	8.550	14.500
	1	0.200	35.000	17.600	34.800

b)   **Character Boxplot**

```
                    ------
         ----I+   I----      *       *                                    o
                    ------
         +---------+---------+---------+---------+---------+------minutes
        0.0       7.0      14.0      21.0      28.0      35.0
```

The distribution is heavily skewed right.

c)   Because of the heavy skewness, the median best represents the center of the distribution. Its value is M= 3.5.

7

Confidence Interval Estimation

A point estimator of a parameter is a single statistic that is subject to variability because its value depends on the particular sample that is taken from the population. Consequently the estimate may assume a value that is somewhat removed from the true value of the parameter. A *confidence interval estimate*, on the other hand, gives an interval of values that should contain the true value of the parameter with a high degree of confidence.

Important issues—Chapter 7

- The *margin of error* is the maximum error that one would expect of the estimate for a specified confidence level.

- Three things are needed in order to develop a confidence interval:

  * A good point estimator of the parameter,
  * The sampling distribution (or approximate sampling distribution) of the point estimator, and
  * The desired confidence level, that is usually stated as a percent.

- We call z^* the *upper critical value* for the standard normal distribution. Following are a few values for z^* that are obtained from the standard normal probability table.

  ```
  1 - α     .80     .85     .90     .95     .98     .99
  z*       1.28    1.44    1.645   1.96    2.33    2.58
  ```

- The usefulness of a confidence interval is judged by its validity and precision. Validity is measured by the confidence level, which is the probability that the interval will contain the true value of the parameter. Precision is measured by the length of the interval.

- The large-sample confidence interval for a population proportion is given by $p \pm z^* \sqrt{p(1-p)/n}$.

- If the distribution is normal, near normal or, symmetrical with tails that are not excessively long, the confidence interval for the mean should be based on $\overline{y}$. When the population standard deviation, σ, is known it takes the form $\overline{y} \pm z^* \sigma/\sqrt{n}$. If the population standard deviation is unknown it takes the form $\overline{y} \pm t^* s/\sqrt{n}$.

- If the distribution is skewed, the best measure of center is the median and therefore the confidence interval should be for the population median. The confidence interval is formed by two sample observations, (C_L, C_H) where C_L is found by counting in from the low end of an ordered stem and leaf plot and C_H is found by counting in from the high end of the ordered stem and leaf plot. The distance we count in is given by the Location of C. For large samples, Location of C = $[n - z^*\sqrt{n}]/2$. For small samples the Location of C is found in Table 7.3.

- If the distribution is symmetric with long tails and the sample size is large, then the confidence interval for the mean should be based on $\bar{y}_T$. It takes the form $\bar{y}_T \pm z^* s_T/\sqrt{k}$

- If the distribution is symmetric with long tails and the sample size is small it is recommended that the confidence interval be for the population median.

Glossary

A **confidence interval** for a population parameter is an interval of possible values for the unknown parameter. The interval is computed from sample data in such a way that we have a high degree of confidence that the interval contains the true value of the parameter. The confidence, stated as a percent, is the **confidence level**.

The **standard error of a statistic** is the standard deviation of its sampling distribution when all unknown population parameters have been estimated.

A $(1-\pi)100\%$ **confidence interval for π** is given by the limits $p \pm z^*\sqrt{p(1-p)/n}$

The **sample size required to estimate π** with a $(1 - \alpha)100\%$ confidence interval so that the bound on the margin of error is B is $\quad n = [z^*\sqrt{p(1-p)/B}]^2$

The z-interval: A $(1-\alpha)100\%$ confidence interval for μ based on $\bar{y}$ is given by the limits $\bar{y} \pm z^*\sigma/\sqrt{n}$.

The **sample size required to estimate μ** with a $(1 - \alpha)100\%$ confidence interval so that the bound on the margin of error is B is $\quad n = (z^*\sigma/B)^2$

The t-interval: A $(1-\alpha)100\%$ confidence interval for μ is given by the limits $\bar{y} \pm t^* s/\sqrt{n}$ where t^* is the upper $\alpha/2$ critical value found in the t-table with degrees of freedom = n-1.

The **winsorized sample** is obtained by replacing the trimmed values with the values that were next in line for trimming.

The **standard deviation of a trimmed sample** based on the Winsorized sample is given by $s_T = s_W\sqrt{(n-1)/(k-1)}$ where s_W is the ordinary sample standard deviation of the Winsorized sample and k is the size of the trimmed sample.

The **standard error of the trimmed mean,** $\bar{y}_T$, is given by $SE(\bar{y}_T) = s_T/\sqrt{k}$ where s_T is the standard deviation of the trimmed sample and k is the size of the trimmed sample.

A **large sample (1-α)100% confidence interval for μ based on** $\bar{y}_T$ is given by the limits $\bar{y}_T \pm z^* s_T/\sqrt{k}$ where s_T is the standard deviation of the trimmed sample based on the Winsorized sample and k is the size of the trimmed sample.

Practice Test 1

I. Multiple Choice

1. The percent of the senior class that makes the honor roll is 15% ± 3%. The 15%± 3% is a
a) parameter.
b) statistic.
c) point estimate.
d) confidence interval estimate.

2. The median age at which women get married for the first time is 24.8 years ± 6.5 years. The 6.5 is
a) a parameter.
b) the standard error.
c) the margin of error.
d) a critical value.

3. If the parent distribution is skewed right then a confidence interval for the center of the distribution should be for the parameter
a) μ b) θ c) M   d) π

4. Suppose a random sample of size 100 from a population with tails that are not unusually long produced a mean of 265.3 and a standard deviation of 47.8. A 99% confidence interval for the population mean is
a) 265.3 ± 9.37
b) 265.3 ± 11.14
c) 265.3 ± 4.78
d) 265.3 ± 12.33

5. A random sample of size 400 produced a sample proportion of .47. A 95% confidence interval for the population proportion is
a) $.47 \pm .025$
b) $.47 \pm .049$
c) $.47 \pm .098$
d) $.47 \pm .058$

6. If the parent distribution is skewed then a confidence interval for the center of the distribution should be based on the statistic
a) $\bar{y}$ b) $\bar{y}_T$ c) M   d) s

7. Suppose a random sample of size 250 from a population with tails that are not unusually long produced a mean of 36.4 and a standard deviation of 7.3. A 98% confidence interval for the population mean is
a) $36.4 \pm .905$
b) $36.4 \pm .462$
c) $36.4 \pm .759$
d) 36.4 ± 1.076

II. Problems

8. A criminologist is studying the entry age of criminals in prisons around the U.S. Identify the parent population, a parameter of interest, and an associated statistic that might estimate the parameter. Give the form of a 95% confidence interval for the parameter.

9. A political scientist is studying the percent of registered voters in major cities with populations in excess of 200,000 who voted in the last presidential election. Identify the parent population, a parameter of interest, and an associated statistic that might estimate the parameter. Give the form of a 95% confidence interval for the parameter.

10. A sawmill operator is interested in the amount of timber in a given region. Identify the parent population, a parameter of interest, and an associated statistic that might estimate the parameter. Give the form of a 95% confidence interval for the parameter.

11. During a recent "smoke-out" a random sample of 200 smokers indicated that 38 had quit smoking. Find a 95% confidence interval for the true proportion of all smokers who quit smoking during the "smoke-out".

12. If in exercise 7 we wanted to estimate within 4% with 95% confidence the percent of smokers who quit, how large of a sample would be needed?

13. A new gasoline additive is reported to increase gas mileage by at least 3 miles per gallon. A consumer agent measures the miles per gallon on 12 different makes of autos both with and without the additive. The difference in miles per gallon are listed below:

1.8, 2.4, 4.1, 2.7, 3.2, 1.2, 2.6, 0.5, 1.6, 2.1, 0.9, 3.3

Assuming the scores are from a normal population, obtain a 98% confidence interval for the mean increase in gas mileage.

14. Following are scores of 35 workers on a job performance rating scale:

23.6, 21.8, 19.6, 23.3, 18.4, 19.7, 24.3, 25.0, 23.1, 22.2, 24.6, 31.6, 22.7, 24.8, 25.4, 22.7, 23.9, 14.2, 3.7, 20.4, 27.6, 29.4, 26.4, 24.1, 21.5, 16.3, 24.3, 25.7, 26.4, 21.5, 37.2, 20.3, 23.7, 23.5, 22.9

a) Construct an ordered stem and leaf plot.
b) Construct a letter value display.
c) Draw a box plot and comment on the distributional shape.
d) What parameter best describes a typical (or representative) score?
e) Construct a 95% confidence interval for the above parameter.

15. How large of a sample is needed to estimate a population mean with a 95% confidence interval to within 2 units if the population standard deviation is 15?

Answers to Practice Test 1

1. d 2. c 3. b 4. d 5. b 6. c 7. d

8. parent population—criminals in U. S. prisons, parameter—population mean, statistic— $\bar{y}$
    confidence interval $\bar{y} \pm t^* s/\sqrt{n}$

9. parent population— of registered voters in major cities with populations in excess of 200,000 parameter—population proportion, statistic—p , confidence interval $p \pm z^* \sqrt{p(1-p)/n}$

10. parent population— timber in the region, parameter—population mean, statistic— $\bar{y}$
    confidence interval $\bar{y} \pm t^* s/\sqrt{n}$

11. (.136, .244)   12. 370

13. (1.37, 3.03)

14. a)
```
    Stem-and-leaf of rating    N = 35
        Leaf Unit = 1.0

     1     1 4
     2     1 6
     5     1 899
    10     2 00111
   (10)    2 2222333333
    15     2 44444555
     7     2 667
     4     2 9
     3     3 1
     2     3 3
     1     3
     1     3 7
```

104

b) **Letter Value Display**

	DEPTH	LOWER	UPPER	MID	SPREAD
N=	35				
M	18.0		23.600	23.600	
H	9.5	21.650	25.200	23.425	3.550
E	5.0	19.700	27.600	23.650	7.900
D	3.0	18.400	31.600	25.000	13.200
C	2.0	16.300	33.700	25.000	17.400
	1	14.200	37.200	25.700	23.000

c) **Character Boxplot**

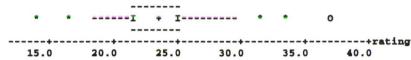

The distribution is long-tailed.

d) trimmed population mean
e) (22.43, 24.64)

15. 217

Practice Test 2

I. Multiple Choice

1. The percent of new TV shows that are renewed is 35% ± 8%. The 35% is
a) a parameter.
b) a point estimate.
c) an estimator.
d) a confidence interval estimate.

2. The mortality rate for a certain disease last year was 30%. To find a confidence interval estimate of the overall mortality rate we would need
a) the confidence level.
b) the sample size.
c) the confidence level and sample size.
d) none of the above.

3. If the parent distribution is normal then a confidence interval for the population mean should be based on the statistic
a) $\bar{y}$ b) $\bar{y}_T$ c) M   d) s

4. If the parent distribution is symmetrical with long tails then a confidence interval for the population mean should be based on the statistic
a) $\bar{y}$ b) $\bar{y}_T$ c) M   d) s

105

5. Suppose a random sample of size 100 from a population that is symmetrical with long tails produced a 10% trimmed mean of 78.5 and a trimmed standard deviation of 15.2. A 95% confidence interval for the population mean is
a) 78.5 ± 2.98
b) 78.5 ± 3.33
c) 78.5 ± 2.50
d) 78.5 ± 2.80

6. A random sample of size 200 produced a sample proportion of .65. A 99% confidence interval for the population proportion is
a) $.65 \pm .087$
b) $.65 \pm .091$
c) $.65 \pm .066$
d) $.65 \pm .069$

II. Problems

7. A sociologist is studying income levels in a particular county of the state. A random sample of 50 families is selected and their gross incomes recorded. Identify the parent population, a parameter of interest, and an associated statistic that might estimate the parameter. Give the form of a 95% confidence interval for the parameter.

8. A chemist is analyzing the concentration of copper in a hair sample taken from an individual. Identify the parent population, a parameter of interest, and an associated statistic that might estimate the parameter. Give the form of a 95% confidence interval for the parameter.

9. A survey of 70 randomly selected households showed that the average monthly cost of electricity was $84.38 and the standard deviation was $24.82. Find a 99% confidence interval for the mean cost of electricity for homes in this area.

10. Acquisition refers to the stage during which a new response is learned and is gradually strengthened. The following data, obtained from a learning experiment, represent the rate of acquisition as measured by the number of trials it took each of twenty subjects to successfully perform a memory task.

12 10 16 27 18 13 14 20 8 23 33 13 14 38 19 6 11 15 10 42

a) Construct an ordered stem and leaf plot.
b) Construct a letter value display.
c) Draw a box plot and comment on the distributional shape.
d) What parameter best describes a typical (or representative) score.
e) Construct a 95% confidence interval for the above parameter.

11. Out of 45 individuals, 18 were able to master a task after being given a drug. Find a 98% confidence interval for the true proportion that would be able to complete the task after the drug.

12. If in exercise 11 we wanted to estimate within 1% with 98% confidence, how large of a sample would be needed?

13. A college official is interested in estimating the mean grade point average for freshmen. It has been determined that $\sigma = 1.0$. Determine the number of student records to be selected in order to estimate the mean grade point average using a 98% confidence interval in which the length of the interval is only .2.

14. An employee of the state highway patrol desires to estimate the proportion of drivers who wear seat belts. He randomly selects 100 drivers and finds that 62 wear seat belts on a regular basis. Obtain a 90% confidence interval for the true proportion of drivers who wear seat belts.

15. A survey of 50 randomly selected households showed that the average cost of preparing a Thanksgiving dinner to be $18.23 with a standard deviation of $4.82. Find a 99% confidence interval for the mean cost of preparing a Thanksgiving meal.

16. The Food and Drug Administration (FDA) requires manufacturers of food products to list FDA estimates of the contents of their products. Suppose the FDA is certain that the sugar content of a box of cereal is between 1 and 4 ounces but wishes to estimate the mean sugar content to within .1 of an ounce of its true value with a 95% confidence interval. How large of a sample is required?

Answers to Practice Test 2

1. b 2. c 3. a 4. b 5. b 6. a

7. parent population— income levels in a particular county of the state, parameter—population mean, statistic— $\bar{y}$, confidence interval $\bar{y} \pm t^* s/\sqrt{n}$

8. parent population— concentration of copper in a hair sample, parameter—population mean, statistic— $\bar{y}$, confidence interval $\bar{y} \pm t^* s/\sqrt{n}$

9. ($76.73, $92.03)

10. a)
```
Stem-and-leaf of rate        N  = 20
      Leaf Unit = 1.0

  2      0 68
 10      1 00123344
 10      1 5689
  6      2 03
  4      2 7
  3      3 3
  2      3 8
  1      4 2
```

b) Letter Value Display

	DEPTH	LOWER	UPPER	MID	SPREAD
N=	20				
M	10.5		14.500	14.500	
H	5.5	11.500	21.500	16.500	10.000
E	3.0	10.000	33.000	21.500	23.000
D	2.0	8.000	38.000	23.000	30.000
	1	6.000	42.000	24.000	36.000

107

c)   Character Boxplot

```
                      -----------------
             --------I     +          I----------------          *         *
                      -----------------
             ----+---------+---------+---------+---------+---------+--rate
                7.0       14.0      21.0      28.0      35.0      42.0
```
   The distribution is skewed right.
d) median
e) (12, 20)

11. (.23, .57)   12. 13030   13. 543
14. (.54, .70)   15. (16.47, 19.99)   16. 97

8

Hypothesis Testing

Hypothesis testing is an area of statistical inference in which we evaluate a conjecture (which we will call a hypothesis) about some characteristic of the parent population. Usually the hypothesis concerns one of the unknown parameters of the population.

Important issues—Chapter 8

- The *null hypothesis* is the hypothesis being tested in a *test of hypothesis*. The *alternative hypothesis* is the hypothesis believed by the researcher to be true, and is also called the *research hypothesis*.

- A *type I error* is rejecting a true null hypothesis. A *type II error* is failing to reject a false null hypothesis. The probability of a type I error is denoted by α and is called the *level of significance*. It reflects the *size* of the rejection region which determines how small the p-value should be before we reject the null hypothesis. The probability of a type II error is denoted by β.

- A *test statistic* is the statistic that is used to test the null hypothesis. The null hypothesis will be rejected if the test statistic falls in the rejection region.

- A *right-tailed test* is a test of hypothesis where the rejection region is on the right tail of the sampling distribution of the test statistic. A *left-tailed test* is a test where the rejection region is on the left tail of the sampling distribution of the test statistic. A *two-tailed test* is one where the rejection region is on both tails of the sampling distribution of the test statistic.

- The *p-value* is the probability of observing a value of the test statistic at least as extreme as that given by the observed data under the assumption that the null hypothesis is true. The test statistic will fall in the rejection region if and only if the p-value is smaller than the level of significance. In this case, the null hypothesis is rejected.

- The steps to follow in testing hypothesis are:
  * Formulate the null and alternate hypotheses.
  * Decide on an appropriate test statistic.
  * Determine the sampling distribution of the test statistic under the assumption that the null hypothesis is true.

* Collect the data and calculate the p-value and determine whether it is sufficiently small to reject the null hypothesis.
* Interpret the results in a way that a nonstatistician can understand.

- The test of a population mean can be based on either $\bar{y}$ or $\bar{y}_T$. The test is based on $\bar{y}$ if the tails of the parent distribution are not excessively long. It is based on $\bar{y}_T$ if the parent distribution is symmetric with long tails.

- If the parent distribution is skewed, the inference should be concerned with the population median, and the test based on the sample median.

- The test of μ based on $\bar{y}$ is either a z-test or a t-test depending on whether we know the population standard deviation.

- The test of a population proportion is a z-test and is similar to the test of the population mean based on $\bar{y}$.

Glossary

The statement being tested in a test of hypothesis is called the **null hypothesis** and is denoted as H_0.

The **alternative hypothesis,** denoted as H_a, is what is believed to be true if the null hypothesis is false. Usually the person conducting the research wishes to establish that there is a difference between the parameter and the value being tested and thus the alternative is also called the **research hypothesis.**

A **test statistic** is a statistic, calculated from the sample data, which is used to test the hypothesis.

The **rejection region** is those values of the test statistic that lead to the rejection of the null hypothesis.

The **p-value** is the probability (computed when H_0 is assumed to be true) of observing a value of the test statistic at least as extreme as that given by the actual observed data. The smaller the p-value, the stronger is the evidence against the null hypothesis.

Criterion for Rejection of H_0
- If p-value $> .10$ then fail to reject H_0 and declare the results insignificant.
- If $.05 <$ p-value $\leq .10$ you may reject H_0, but the results are only mildly significant.
- If $.01 <$ p-value $\leq .05$ then reject H_0 and declare the results significant.
- If p-value $\leq .01$ then reject H_0 and declare the results highly significant.

A **type I error** is rejecting a null hypothesis that is true. The probability of committing a Type I error is denoted as α.

A **type II error** is failing to reject a null hypothesis that is false. The probability of committing a Type II error is denoted as β.

Equivalence of Confidence Intervals and Two-tailed Tests
A two-tailed test of the null hypothesis of a parameter is rejected at an α level of significance if and only if the hypothesized value of the parameter falls outside a $(1 - \alpha)100\%$ confidence interval for the parameter.

Large Sample Test Of A Population Proportion
Application: Bernoulli Population
Assumption: $n > 30$

Left-Tailed Test	Right-Tailed Test	Two-Tailed Test
$H_0: \pi \geq \pi_0$	$H_0: \pi \leq \pi_0$	$H_0: \pi = \pi_0$
$H_a: \pi < \pi_0$	$H_a: \pi > \pi_0$	$H_a: \pi \neq \pi_0$

Standardized Test Statistic:

$$z = \frac{p - \pi_0}{\sqrt{\pi_0(1-\pi_0)/n}}$$

Test Of μ based on $\bar{y}$
Application: Symmetric population distributions whose tails are not unusually long. If the sample size is small then the parent distribution should not deviate substantially from normality.

Left-Tailed Test	Right-Tailed Test	Two-Tailed Test
$H_0: \mu \geq \mu_0$	$H_0: \mu \leq \mu_0$	$H_0: \mu = \mu_0$
$H_a: \mu < \mu_0$	$H_a: \mu > \mu_0$	$H_a: \mu \neq \mu_0$

Standardized Test Statistic when σ is unknown:

$$t = \frac{\bar{y} - \mu_0}{s/\sqrt{n}}$$

Standardized Test Statistic when σ is known:

$$z = \frac{\bar{y} - \mu_0}{\sigma/\sqrt{n}}$$

Test Of The Population Median

Application: Mainly skewed distributions but can be used to test the median of any continuous distribution.

Left-Tailed Test   Right-Tailed Test   Two-Tailed Test

$H_0: \theta \geq \theta_0$ $H_0: \theta \leq \theta_0$ $H_0: \theta = \theta_0$

$H_a: \theta < \theta_0$ $H_a: \theta > \theta_0$ $H_a: \theta \neq \theta_0$

Small sample: Parent distribution is continuous and $n \leq 20$
Test Statistic:

$$T = \text{no. of observations} > \theta_0$$

Large Sample : Parent distribution is continuous and $n > 20$
Test Statistic:

$$z = \frac{2T - n}{\sqrt{n}}$$

Large Sample Test of μ Based on $\overline{y}_T$

Application: Symmetric population distribution with long tails
Assumption: $n > 30$

Left-Tailed Test   Right-Tailed Test   Two-Tailed Test

$H_0: \mu \geq \mu_0$ $H_0: \mu \leq \mu_0$ $H_0: \mu = \mu_0$

$H_a: \mu < \mu_0$ $H_a: \mu > \mu_0$ $H_a: \mu \neq \mu_0$

Standardized Test Statistic:

$$z = \frac{\overline{y}_T - \mu_0}{s_T / \sqrt{k}}$$

where s_T is the standard deviation of the trimmed sample based on the Winsorized sample and k is the size of the trimmed sample.

Practice Test 1

I. Multiple Choice

1. The p-value for a two-tailed test is
a) computed just as in a one-tailed test.
b) computed just as in a one-tailed test and then halved.
c) computed just as in a one-tailed test and then doubled.
d) not at all computed as in a one-tailed test.

2. Which of the following p-values would lead to rejection of the null hypothesis?
a) p-value = 15%
b) p-value = 95%
c) p-value = 100%
d) p-value = 1%

3 In testing H_0: $\mu = 10$ vs H_a: $\mu > 10$ with n = 100 , $\bar{y} = 11.1$, and s = 4.6 we have
a) t = 2.39 with the p-value found in the t-table.
b) z = 2.39 with the p-value found in the z-table.
c) t = -2.39 with the p-value found in the t-table.
d) z = -2.39 with the p-value found in the z-table.

4 If it appears that the parent distribution is symmetrical with long tails then a test of æ should be based on
a) $\bar{y}$
b) $\bar{y}_T$
c) M
d) p

5 In testing H_0: $\mu = 50$ vs H_a: $\mu > 50$ with n = 100 , $\bar{y} = 52.4$ which of the following values of s would lead to rejection of H_0?
a) 200
b) 20
c) 2
d) none of the above.

6. A manufacturer would like to see if their product has a greater shelf-life than their competitor. If μ_1 denotes the mean shelf-life of the manufacturer's product and μ_2 denotes the mean shelf-life of the competitor then the research hypothesis is
a) $\mu_1 < \mu_2$
b) $\mu_1 \leq \mu_2$
c) $\mu_1 > \mu_2$
d) $\mu_1 \geq \mu_2$

II. Problems

7. It has been reported that at least 10% of the adult male population suffers from kidney stones. A random sample of 900 males is selected and tested for kidney stones. If only 67 in the sample suffer from kidney stones, is there statistical evidence to deny the above claim?

8. Do the following data that have been organized in an ordered stem and leaf plot indicate that the population median exceeds 50?

```
1|6
2|0 2 3 5 8
3|1 2 4 6 7 9
4|1 3 5 7
5|0 4 8
6|1 5
7|2
8|5
9|1
```

9. A random sample of 150 students revealed that 84 had their own car on campus. Is there statistical evidence that more than half the student body has their own car on campus?

10. Ten high school seniors took the ACT test and made the following scores:

28 26 30 24 25 29 31 26 23 28

Past ACT scores at their high school have shown the scores to be normally distributed with a mean of 25. Is there evidence to suggest that this year's class is above average? Perform a test of hypothesis to include the p-value and give an interpretation.

11. An employee of a large company feels that the coffee dispensing machine in their office is not filling the cups properly. She contends that the median amount put in a 12 ounce cup is less than 11.8 ounces. From the following sample of 24 cups evaluate her claim:

11.4, 11.7, 11.1, 11.7, 11.9, 11.6, 11.3, 11.7, 11.8, 11.6,
11.9, 11.4, 11.3, 11.5, 11.7, 11.8, 11.7, 11.8, 11.5, 11.2,
11.9, 11.8, 11.6, 11.77.

12. Test the following hypothesis using the summary data that is provided:

$H_0: \mu \geq 70$ $H_a: \mu < 70$

$n = 50$ $\bar{y}_{T.10} = 68.3$ $s_T = 4.7$

13. The county agent believes that more than 40% of the county acreage is in need of lime. Out of 60 random soil samples it was found that 28 should a need for lime. Does this data support the agents claim?

14. A city official claims that the average city tax is no more than $120 per year. A random sample of 100 residents found an average of $134.60 with a standard deviation of $37.40. Does this cast doubt on the official's claim?

Answers to Practice Test 1

1. c 2. d 3. a 4. b 5. c 6. c
7. Yes, $z = -2.56$, p-value = .0052
8. $T = 7$, $n = 23$, $z = -1.88$, p-value = .0301. There is sufficient evidence that the population median exceeds 50.
9. $z = 1.47$, p-value = .0708. There is mildly significant evidence that more than half of the student body has their own car on campus.
10. $t = 2.41$, $.01 <$ p-value $< .025$. There is significant evidence that this year's class is above average.
11. $T = 3$, $n = 24$, $z = -3.67$, p-value $< .001$. There is highly significant evidence that the median contents is less than 11.8 ounces.
12. $z = -2.56$, p-value = .0052. Reject H_0. There is significant evidence that $\mu < 70$.
13. $z = 1.06$, p-value = .1446. There is insufficient evidence to support the agent's claim.
14. $t = 3.90$, p-value $< .0002$. There is highly significant evidence that the city official is in error.

Practice Test 2

I. Multiple Choice

1. In testing hypothesis the results would be judged highly significant if the
a) p-value = 5%
b) p-value = 99%
c) p-value = 10%
d) p-value = 1%

2. Which of the following p-values would lead to rejection of the null hypothesis?
a) p-value = 2%
b) p-value = 98%
c) p-value = 100%
d) p-value = 12%

3. Suppose μ is the mean of a population that is normally distributed. In testing
H_0: $\mu = 100$ vs H_a: $\mu < 100$ with $n = 20$, $\bar{y} = 97.3$, and $s = 12.7$ we have
a) $t = 1.303$ with the p-value found in the t-table.
b) $z = 1.303$ with the p-value found in the z-table.

c) t = -1.303 with the p-value found in the t-table.
d) z = -1.303 with the p-value found in the z-table.

4. If it appears that the parent distribution is skewed right then a test of hypothesis about the center of the distribution should be based on
a) the sample mean
b) a trimmed sample mean
c) the sample proportion
d) the sample median

5. In testing H_0: $\mu = 25$ vs H_a : $\mu > 25$ with n = 100 , $\bar{y}$ = 27.4 which of the following values of s would lead to rejection of H_0?
a) 25
b) 2.5
c) 250
d) none of the above.

II. Problems

6. A sample of 90 observations yielded a) $\bar{y}_T$ = 73.6 and s_T = 16.9. Use this information to test
H_0 : $\mu \geq 75$ vs H_a : $\mu < 75$
Be sure to calculate the p-value and interpret the results.

7. A supplier of auto parts believes that he can produce a particular part and ship it for under $28.00 per part. A sample of 50 such items were shipped for an average cost of $26.94 with a standard deviation of $3.43. Is there statistical evidence that supports the suppliers claim?

8. The parts supplied by the supplier in Exercise 7 can be defective. Their claim is that no more than 8% of the parts are defective. A random sample of 120 parts revealed that 15 were defective. Does this cast doubt on the supplier's claim?

9. The production cost to produce a certain item is $120. A sample of 35 items produced by a new process yielded an average of $112.65 and a standard deviation of $28.72. Is there statistical evidence that the production cost is less with the new process?

10. An environmentalist suspects that a chemical plant is violating EPA standards for nitrogen by-product concentration in their waste that is being dumped in a major river. The EPA standard says that the concentration should be below 120 ppm. Test were taken twice a month for a year to determine the concentration of nitrogenous waste being put in the river. Assuming the parent distribution appears normal in shape, do the following data indicate that the chemical plant's concentration exceeds the EPA standard?

 120.5 130.1 120.3 125.3 124.0 121.3 118.6 119.3 123.9 120.5
 117.6 125.2 121.8 131.0 128.5 117.4 119.9 123.2 122.0 118.5

11. It is believed that 30% of the adult population has never flown in an airplane. A random sample of 2000 adults found that 1500 had flown at least once. Does this data cast doubt on the above claim?

12. Fifteen third graders took an achievement test and made the following scores:

34 28 31 26 36 23 29 27 37 26 22 27 39 25 29

Past achievement test scores have been normally distributed with a mean of 25. Is there evidence to suggest that this year's class is above average? Perform a test of hypothesis to include the p-value and give an interpretation.

13. The operator of a department store feels certain that more than 70% of her customers are teenagers. A random sample of 75 customers revealed that 54 were teenagers. Does the data support her conjecture?

14. A conjecture is made that the median life span of a certain animal held in captivity exceeds 7.8 years. The ages at death of a random sample of 9 such animals were estimated to be

8.4 12.8 8.9 7.1 13.2 10.8 6.3 8.1 9.4
Is there evidence that the median life span exceeds 7.8 years?

15. The temperature of the coolant in a particular area of a nuclear reactor should not be significantly above 250 degrees. Does the following random sample indicate that the standard is being violated?

244 249 254 259 238 247 251 258 244 252 260 259

Answers to Practice Test 2

1. d 2. a 3. c 4. d 5. b
6. $z = -.70$, p-value = .242. There is insufficient evidence to support that $\mu < 75$.
7. $t = -2.19$, p-value < .025. There is sufficient evidence that the supplier can ship for under $28.00.
8. $z = 1.82$, p-value = .0344. There is sufficient evidence to reject the suppliers claim and conclude that significantly more than 8% are defective.
9. $t = -1.51$, $.05 < $ p-value $ < .10$. There is mildly significant evidence that the production cost is less with the new process.
10. $t = 2.756$, $.005 < $ p-value $ < .01$. There is highly significant evidence that the chemical plant's concentration exceeds the EPA standard.
11. $z = 43.92$, p-value < .00001. There is highly significant evidence to reject the claim that 30% has never flown on an airplane.
12. $t = 3.22$, p-value < .01. There is highly significant evidence that this class is above average.
13. $z = .38$, p-value = .352. There is insufficient evidence to support the operator's claim.
14. $T = 7$, p-value = .09. There is mildly significant evidence that the median life span exceeds 7.8 years.
15. $T = 7$, p-value = .387. There is insufficient evidence that the median temperature exceeds 250 degrees.

9

Inference About The Difference Between Two Parameters

A widely used method of statistical analysis is the comparison of two populations. Often we wish to compare a new procedure with an old established procedure, or to compare one product with another, or to compare one treatment with another treatment. The key word is *compare*; we wish to compare the characteristics of one population with those of another population. To compare the two populations, random samples from each must be chosen and then used to make inferences about the parameters of the two populations.

Important issues—Chapter 9

- Three basic inference problems are addressed in this chapter-- the comparison of two population proportions, the comparison of two population centers using independent samples, and the comparison of two population centers using matched samples.

- Only the large sample inference procedures for the difference in two population proportions is given. The confidence interval is given by

$$(p_1 - p_2) \pm z^* \sqrt{p_1(1-p_1)/n_1 + p_2(1-p_2)/n_2}$$

and the test statistic for testing the hypothesis that $\pi_1 = \pi_2$ is

$$z = \frac{p_1 - p_2}{p(1-p)\sqrt{1/n_1 + 1/n_2}}$$

where

$$p = \frac{x_1 + x_2}{n_1 + n_2}$$

is the pooled estimate of π.

- The confidence interval for $\mu_1 - \mu_2$ based on $\bar{y}_1 - \bar{y}_2$ is given by

$$(\bar{y}_1 - \bar{y}_2) \pm t^* \, SE(\bar{y}_1 - \bar{y}_2)$$

- The test statistic for testing $\mu_1 - \mu_2$ is

$$t = \frac{(\bar{y}_1 - \bar{y}_2}{SE(\bar{y}_1 - \bar{y}_2)}$$

- If the population variances are equal then

$$SE(\bar{y}_1 - \bar{y}_2) = s_p \sqrt{1/n_1 + 1/n_2}$$

and

$$t = \frac{(\bar{y}_1 - \bar{y}_2) - (\mu_1 - \mu_2)}{s_p \sqrt{1/n_1 + 1/n_2}}$$

has a Student's t distribution with $n_1 + n_2 - 2$ degrees of freedom.

- If the population variances are unequal then

$$SE(\bar{y}_1 - \bar{y}_2) = \sqrt{s_1^2/n_1 + s_2^2/n_2}.$$

and

$$t = \frac{(\bar{y}_1 - \bar{y}_2) - (\mu_1 - \mu_2)}{s_1^2/n_1 + s_2^2/n_2}.$$

has an approximate t-distribution with degrees of freedom given by

$$df = \frac{[s_1^2/n_1 + s_2^2/n_2]^2}{\frac{(s_1^2/n_1)^2}{n_1 - 1} + \frac{(s_2^2/n_2)^2}{n_2 - 1}}$$

- The *Wilcoxon Rank Sum test* is an alternative to the two-sample t-test when the assumption of normality is not valid. For the Wilcoxon rank sum test the usual t-test is applied to the rank transformed data.

- When the samples are matched, the test of the difference in population means is called the *matched-pairs t-test*. The test statistic is

$$t = \frac{\overline{d}}{s_d/\sqrt{n}}$$

  which has a Student's t distribution with n-1 degrees of freedom. Again, the assumption of normality is required for this t-test unless the sample size is large. If the assumption is not met then the *Wilcoxon Signed Rank Test* is recommended. Here the usual t-test is applied to the signed ranks.

- Large sample inference procedures for μ_1-μ_2 based on $\overline{y}_{T1}$- $\overline{y}_{T2}$ are the same as those given above except that the ordinary sample means and standard deviations are replaced with the trimmed means and trimmed standard deviations.

Glossary

Inference for π_1 - π_2

Application: Bernoulli populations

Assumptions: Independent samples and $n_1 > 30$, $n_2 > 30$

A (1- α)100% confidence interval for π_1 - π_2 is given by the limits

$$(p_1 - p_2) \pm z^* \sqrt{p_1(1-p_1)/n_1 + p_2(1-p_2)/n_2}$$

Left-tailed Test	Right-tailed Test	Two-tailed Test
H_0: π_1-$\pi_2 \geq 0$	H_0: π_1-$\pi_2 \leq 0$	H_0: π_1-$\pi_2 = 0$
H_a: π_1-$\pi_2 < 0$	H_a: π_1-$\pi_2 > 0$	H_a: π_1-$\pi_2 \neq 0$

Standardized Test Statistic:

$$z = \frac{p_1 - p_2}{\sqrt{p(1-p)}\sqrt{1/n_1 + 1/n_2}}$$

where

$$p_1 = \frac{x_1}{n_1}, \quad p_2 = \frac{x_2}{n_2}, \quad \text{and } p = \frac{x_1 + x_2}{n_1 + n_2}$$

and x_1 is the number of successes in sample 1 and x_2 is the number of successes in sample 2.

Inference for $\mu_1 - \mu_2$ Based on $\bar{y}_1 - \bar{y}_2$

Application: Symmetric distributions whose tails are not excessively long. If the parent distributions deviate substantially from normality then the sample sizes should be larger than 30.

Assumption: The two samples are independent of each other and the population variances are unequal.

Estimator: $\bar{y}_1 - \bar{y}_2$

Standard Error: $SE(\bar{y}_1 - \bar{y}_2) = \sqrt{s_1^2/n_1 + s_2^2/n_2}$

A $(1-\alpha)100\%$ confidence interval for $\mu_1 - \mu_2$ is given by the limits- $(\bar{y}_1 - \bar{y}_2) \pm t^* \sqrt{s_1^2/n_1 + s_2^2/n_2}$

Left-Tailed Test	Right-Tailed Test	Two-Tailed Test
$H_0: \mu_1 - \mu_2 \geq 0$	$H_0: \mu_1 - \mu_2 \leq 0$	$H_0: \mu_1 - \mu_2 = 0$
$H_a: \mu_1 - \mu_2 < 0$	$H_a: \mu_1 - \mu_2 > 0$	$H_a: \mu_1 - \mu_2 \neq 0$

Standardized Test Statistic:

$$t = \frac{(\bar{y}_1 - \bar{y}_2)}{\sqrt{s_1^2/n_1 + s_2^2/n_2}}$$

t^* for the confidence interval and the p-value of the test of hypothesis are found in the t-table with degrees of freedom given by

$$df = \frac{[s_1^2/n_1 + s_2^2/n_2]^2}{\dfrac{(s_1^2/n_1)^2}{n_1 - 1} + \dfrac{(s_2^2/n_2)^2}{n_2 - 1}} \quad \text{(round down to nearest integer)}$$

Pooled T-Procedures Based On $\bar{y}_1 - \bar{y}_2$

Application: Symmetric distributions whose tails are not excessively long. If the parent distributions deviate substantially from normality then the sample sizes should be larger than 30.

Assumptions: The two samples are independent of each other and the population variances are equal.

Confidence interval: A $(1-\alpha)100\%$ confidence interval for $\mu_1 - \mu_2$ is given by the limits:

$$(\bar{y}_1 - \bar{y}_2) \pm t^* s_p \sqrt{1/n_1 + 1/n_2}$$

where

$$s_p = \sqrt{\frac{(n_1-1)s_1^2 + (n_2-1)s_2^2}{n_1 + n_2 - 2}}$$

The test of hypothesis is called the **Pooled t-test**:

Left-Tail Test	Right-Tailed Test	Two-Tailed Test
H_0: $\mu_1 - \mu_2 \geq 0$	H_0: $\mu_1 - \mu_2 \leq 0$	H_0: $\mu_1 - \mu_2 = 0$
H_a: $\mu_1 - \mu_2 < 0$	H_a: $\mu_1 - \mu_2 > 0$	H_a: $\mu_1 - \mu_2 \neq 0$

Standardized Test Statistic:

$$t = \frac{\overline{y}_1 - \overline{y}_2}{s_p\sqrt{1/n_1 + 1/n_2}}$$

t^* for the confidence interval and the p-value for the test of hypothesis are obtained from the t distribution with $n_1 + n_2 - 2$ degrees of freedom.

The Two-Sample Wilcoxon Rank Sum Test

Application: Comparing locations of general populations.

Assumptions: Population distributions are similar except for possibly different centers. Samples are independent. $n_1 \geq 10$ and $n_2 \geq 10$.

The test of hypothesis procedures are the same as the pooled t-test applied to the rank-transformed data.

Inferences For $\mu_1 - \mu_2$ Based On Matched Samples

Application: Matched samples.

Assumptions: If the sample size is small then the distribution of difference scores should be approximately normally distributed.

Confidence interval: A $(1-\alpha)100\%$ confidence interval for $\mu_1 - \mu_2$ based on matched samples is

$$\overline{d} \pm t^* s_d/\sqrt{n}$$

Matched-pairs t-test

Left-Tailed Test	Right-Tailed Test	Two-Tailed Test
H_0: $\mu_d \geq 0$	H_0: $\mu_d \leq 0$	H_0: $\mu_d = 0$
H_a: $\mu_d < 0$	H_a: $\mu_d > 0$	H_a: $\mu_d \neq 0$

Standardized Test Statistic:

$$t = \frac{\overline{d}}{s_d/\sqrt{n}}$$

t^* for the confidence interval and the p-value for the test of hypothesis are obtained from the t distribution with $n - 1$ degrees of freedom.

Large Sample Inference for $\mu_1 - \mu_2$ based on $\bar{y}_{T1} - \bar{y}_{T2}$

Application: Symmetric distributions with long tails
Assumptions: The two samples are independent of each other, $n_1 > 30$ and $n_2 > 30$.
Estimator: $\bar{y}_{T1} - \bar{y}_{T2}$

Standard Error: $SE(\bar{y}_{T1} - \bar{y}_{T2}) = \sqrt{s_{T1}^2/k_1 + s_{T2}^2/k_2}$

Confidence interval: A $(1 - \alpha)100\%$ confidence interval for $\mu_1 - \mu_2$ is given by the limits

$$(\bar{y}_{T1} - \bar{y}_{T2}) \pm z^* SE(\bar{y}_{T1} - \bar{y}_{T2})$$

Test of hypothesis:

Left-Tailed Test	Right-Tailed Test	Two-Tailed Test
H_0: $\mu_1 - \mu_2 \geq 0$	H_0: $\mu_1 - \mu_2 \leq 0$	H_0: $\mu_1 - \mu_2 = 0$
H_a: $\mu_1 - \mu_2 < 0$	H_a: $\mu_1 - \mu_2 > 0$	H_a: $\mu_1 - \mu_2 \neq 0$

Standardized Test Statistic:

$$z = \frac{(\bar{y}_{T1} - \bar{y}_{T2})}{\sqrt{s_{T1}^2/k_1 + s_{T2}^2/k_2}}$$

Choosing the Right Procedure

1. Determine if you have independent or matched samples.
2. In the case of independent samples, analyze the samples to determine the shapes of the two parent distributions. In the case of matched samples, analyze the difference scores to determine the shape of the difference population.
3. Based on the shape(s) of the parent distribution(s) decide if the inference should be for means or medians. In other words, are the parent distributions symmetric or skewed? If they are symmetric, are they normally distributed, symmetric with short tails, or symmetric with long tails?
4. If the inference is for the population medians then the Wilcoxon procedures are to be used.
5. If the inference is for the population means, determine if the standard t-procedures apply. That is, if your sample sizes are small, does the normality and homogeneous variances assumptions seem reasonable? If not apply the Wilcoxon procedure. If the sample sizes are large, examine the length of the tails. For symmetric long-tailed distributions use the z-test based on trimmed means. Otherwise use the t-test based on ordinary means.

Practice Test 1

I.   Multiple Choice

1. The pooled t-test of $\mu_1 - \mu_2$ based on $\bar{y}_1 - \bar{y}_2$ requires
a) homogeneous variances.
b) normally distributed parent populations.
c) independent samples.
d) all of the above.

2. The two sample t-test of equality of population means is applicable when
a) the variances are homogeneous and the parent distributions are normal.
b) the variances are homogeneous and there are no restrictions on the parent distributions.
c) the variances are non-homogeneous and the parent distributions are normal.
d) the variances are non-homogeneous and there are no restrictions on the parent distributions.

3. In a test of hypothesis if the parent distributions appear similarly skewed and the samples are independent then the recommended test is
a) a test based on $\bar{y}_1 - \bar{y}_2$.
b) a test based on $\bar{y}_{T.1} - \bar{y}_{T.2}$.
c) the Wilcoxon rank sum test.
d) the matched pairs t-test.

4. For the matched pairs t-test it is assumed that
a) the samples are independent.
b) the difference scores are normally distributed.
c) the variances are non-homogeneous.
d) all of the above.

5. A manufacturer would like to see if their product has a greater shelf-life than their competitor. If μ_1 denotes the mean shelf-life of the manufacturer's product and μ_2 denotes the mean shelf-life of the competitor then the research hypothesis is
a) $\mu_1 < \mu_2$
b) $\mu_1 \leq \mu_2$
c) $\mu_1 > \mu_2$
d) $\mu_1 \geq \mu_2$

6. Suppose in Exercise 5 the distribution of the shelf-life of the product in question appears close to normal in shape, then the recommended test of the hypothesis is
a) a test based on $\bar{y}_1 - \bar{y}_2$.
b) a test based on $\bar{y}_{T.1} - \bar{y}_{T.2}$.
c) the Wilcoxon rank sum test.
d) the matched pairs t-test.

7. Suppose a journalist wishes to determine if the proportions of male and female adults that read her column are statistically different. If π_1 denotes the proportion of males and π_2 denotes the proportion of females that read her column then the research hypothesis is

a) $\pi_1 < \pi_2$

b) $\pi_1 \leq \pi_2$

c) $\pi_1 > \pi_2$

d) $\pi_1 \neq \pi_2$

8. In Exercise 7 the recommended test is

a) a test based on $\bar{y}_1 - \bar{y}_2$.

b) the Wilcoxon rank sum test.

c) the matched pairs t-test.

d) a test of equality of population proportions.

II. Problems

9. Following is a stroke index (ml/beat/m2) in pre and post treatment studies of coronary circulation in chronic severe anemia.

Case	1	2	3	4	5	6	7	8
Before treatment	89	57	53	57	68	72	51	65
After treatment	56	63	55	40	62	46	59	41

Assuming normality can we conclude that treatment lowers the stroke index?

10. Sixty democrats and 45 republicans were asked if they agreed with the recent action of the local school board. Thirty-five of the democrats and 21 of the republicans disagreed with the school board. Test to see if there is a difference in the proportions of democrats and republicans who disagree with the school board's action.

11. The mean time for a drug to take effect was obtained from a random sample of subjects using a new formulation and from a random sample of subjects using the old formulation.. Estimate from the following data the mean difference in time for the new formulation and the old formulation to take effect. Estimate with a 98% confidence interval.

```
Time to take effect
Old         New
18.6        15.3
13.6        12.4
24.9        19.2
23.4        17.2
27.8        22.5
19.7        20.5
23.5        22.1
26.3        24.7
```

12. A survey of the residents of a community for the elderly showed that 17 of 42 women and 15 of 39 men felt that the living conditions in the community were good to excellent. Is there statistical evidence

that the proportions of men and women who feel that the living conditions are good to excellent are different?

13. To test whether leadership is a trainable quality or not, two groups of subjects were randomly assigned so that one group received special training in leadership and the other received no special instruction. Estimates of the leadership quality of all subjects were as follows:

Leadership course	47	43	36	38	30	22	25	21	14	12	25	39	15
Control group	40	38	42	25	29	26	16	18	8	14	17	33	15

Is there evidence that the course had a positive effect of leadership quality?

14. In a study aimed at determining the effectiveness of a new diet an insurance company selected a sample of 9 overweight men between ages of 45 and 55. Their weight before and after a 60 day diet were recorded as follows:

Before	202	237	173	161	185	210	209	191	200
After	180	221	175	158	180	197	205	196	185

Determine if the diet significantly reduced the weights of the participants.

15. The directors of a small private school decided that they must go co-educational in order to remain open. Contributions from their 1000 club the year before and the year after going co-ed are summarized below.

Before	After
n = 45	n = 41
$\bar{y}_1$ = 2675	$\bar{y}_2$ = 2240
s = 1280	s = 1550

Was there a significance difference in contributions before and after going co-ed?

16. In order to assess the predictive validity of a prognostic rating scale for subjects who have received behavior modification therapy the subjects, following psychotherapy, were separated into an improved or unimproved group.

Prognostic rating scale

Improved group			Unimproved group	
11.9	8.2	6.9	6.6	4.3
11.7	7.4	6.5	5.8	3.9
9.5	7.4	6.3	5.4	3.3
9.4	7.3	6.8	5.1	2.4
8.7	7.1	4.2	5.0	1.7

Is there evidence that the rating scores differ for the two groups?

17. To compare the amount of wear on two types of tires the amount of wear (in millimeters) was measured after being driven 10,000 miles.

```
Tire A  |  6.6   7.0   8.3   8.2   5.2   9.3   7.9   8.5
--------|-----------------------------------------------
Tire B  |  7.4   5.4   8.8   8.0   6.8   9.1   6.3   7.5

Tire A  |  7.8   7.5   6.1   8.9   6.1   9.4   9.1
--------|-----------------------------------------
Tire B  |  7.0   6.6   4.4   7.7   4.2   9.4   9.3
```

Determine if there is a significant difference in the amount of wear for the two makes of tires.

Answers to Practice Test 1

1. d 2. c 3. c 4. b   5. c 6. a   7. d 8. d
9. $t = 2.01$, p-value = .042. There is significant evidence that the treatment lowers the stroke index.
10. $z = 1.18$, p-value = .1190. The difference between the proportions is insignificant.
11. (-2.8, 8.8)
12. $z = .184$, p-value > .10. There is no statistical difference between the proportion of men and women who feel that the living conditions are good to excellent in the community.
13. $t = .80$, p-value = .22. The course did not have a significant positive effect on leadership quality.
14. $t = -2.62$, p-value = .015. The diet significantly reduced the weights of the participants.
15. $t = 1.41$, p-value > .10. There is no significant difference in contributions before and after going co-ed.
16. $t = 4.76$, p-value < .005. There is highly significant evidence that the rating scores differ for the two groups.
17. $t = .99$, p-value > .10. There is no significant difference between the amounts of wear for the two makes of tires.

Practice Test 2

I. Multiple choice

1. A pooled standard deviation is called for in
a) the pooled t-test of $\mu_1 - \mu_2$ based on $\bar{y}_1 - \bar{y}_2$
b) the two-sample t-test based on $\bar{y}_1 - \bar{y}_2$
c) the two-sample Wilcoxon rank sum test.
d) all of the above.

2. The matched pairs t-test
a) assumes that the samples are dependent.
b) assumes that the samples are independent.
c) requires homogeneous variances
d) requires that the absolute value of the differences are ranked.

3. A consumer advocate would like to determine if the median reaction time after taking drug A exceeds the median reaction time after taking drug B. If θ_1 denotes the median reaction time after taking drug A and θ_2 denotes the median reaction time after taking drug B then the research hypothesis is

a) $\theta_1 < \theta_2$
b) $\theta_1 \le \theta_2$
c) $\theta_1 > \theta_2$
d) $\theta_1 \ge \theta_2$

4. Suppose in Exercise 3 the distributions of reaction times in question appear similarly skewed and the samples are independent then the recommended test of hypothesis is
a) a test based on $\bar{y}_1 - \bar{y}_2$.
b) a test based on $\bar{y}_{T.1} - \bar{y}_{T.2}$.
c) the Wilcoxon rank sum test.
d) the matched pairs t-test.

5. In a test of hypothesis if the parent distributions appears symmetrical with long tails and the samples are independent then the recommended test is
a) a test based on $\bar{y}_1 - \bar{y}_2$.
b) a test based on $\bar{y}_{T.1} - \bar{y}_{T.2}$.
c) the Wilcoxon rank sum test.
d) the matched pairs t-test.

6. The Wilcoxon signed rank test assumes that
a) the samples are independent.
b) the difference scores are normally distributed.
c) the difference scores are symmetrical.
d) the variances are homogeneous.

7. If a 95% confidence interval for $\mu_1 - \mu_2$ contains 0 then a test of equality of means would be
a) rejected at the 5% level of significance.
b) rejected at the 95% level of significance.
c) accepted at the 95% level of significance.
d) accepted at the 5% level of significance.

II. Problems

8. Compute the pooled standard deviation from the following two samples

Sample 1: 21 32 16 19 27 22 28 24
Sample 2: 33 28 41 48 37 26 36

9. A new gasoline additive is reported to increase gas mileage. A consumer agent measures the miles per gallon on 12 automobiles with the additive and on 12 automobiles without the additive. The miles per gallon obtained are listed below:

```
With    | 26.3  21.4  17.1  44.3  23.8  19.4  17.6  25.4
--------|--------------------------------------------------
Without | 24.5  17.3  14.7  40.4  22.6  18.9  15.5  24.5

With    | 18.6  21.9  22.6  16.8
--------|--------------------------
Without | 16.5  18.6  21.7  15.6
```

Assuming the scores are from normal populations, test the hypothesis that the additive increased the gas mileage.

10. An automobile insurance company experiences the following claims (in dollars) for each of two makes of cars that had sustained similar damages.

```
Make A  |  353  597  634  696  813  649  593  658
--------|------------------------------------------
Make B  |  453  527  568  498  725  532  568  345
```

Do the claims indicate that one make of car is more expensive to repair?

11. Fifteen (15) employees who have not completed high school were chosen randomly and given a reading test. After the first test they were given formal reading instruction for two weeks. Then a retest of their reading skills was given using an equivalent form of the test. Can we conclude that the reading instruction was beneficial in raising reading scores?

```
Test 1 | 65  72  64  43  55  84  72  52  49  80  38  93  77  62  50
-------|----------------------------------------------------------------
Test 2 | 67  70  72  50  52  86  80  50  62  81  56  90  78  64  58
```

12. An advertising firm wants to compare the proportions of city and county families that shop regularly at the mall. Random samples from each group revealed that of 200 city families 165 and of 120 county families 84 shop regularly at the mall. Is there a significant difference in the proportions from each group that shop at the mall?

13. A consumer is shopping for insurance. She obtained the premiums from two companies on 8 different types of coverage that have been recorded in the following table.

```
Company A  |  138   92  225  175  150  325  260  185
-----------|----------------------------------------------
Company B  |  165  112  207  182  150  345  245  205
```

Do the premiums for the two companies differ significantly?

14. A manufacturer of suntan lotion desired to know if a new ingredient would increase the sun screening ability of their lotion. Seven volunteers had their backs exposed to the sun with the old lotion on one side and the new lotion on the other side. The degree of sunburn was estimated as follows:

```
Volunteer    |  1   2   3   4   5   6   7
-------------|----------------------------
Old lotion   | 42  51  31  61  44  55  48
-------------|----------------------------
New lotion   | 38  53  36  52  33  49  39
```

Using the Wilcoxon matched pairs signed rank test determine if the new ingredient significantly improved the sun screening ability of the sun lotion.

15. In a experiment on the effects of a particular drug on the number of errors in maze-learning behavior of rats, the following data were recorded.

Drug group

$\sum x_i = 324$
$\sum x_i^2 = 6516$
$n = 36$

Control group

$\sum y_i = 256$
$\sum y_i^2 = 4352$
$n = 33$

Determine if the drug significantly increased the maze-learning scores.

16. A random sample of 35 subjects were asked to estimate the height of a tree with one eye closed. The mean was 15.6 feet with a standard deviation of 4.2 feet. A second random sample of 35 subjects were asked to perform the same task with both eyes open. Their mean was 14.2 feet with a standard deviation of 3.8 feet. Do the estimates differ significantly?

17. A researcher wants to determine whether or not a given drug has any effect on the score of human subjects performing a task of psychomotor coordination. Nine subjects in an experimental group received an oral administration of the drug. Ten subjects in a control group received a placebo.

Psychomotor coordination score

```
Experimental  | 12  14  10   8  16   5   3   9  11
--------------|------------------------------------------
Control       | 21  18  14  20  11  19   8  12  13  15
```

Is there a significant difference in the coordination scores for the experimental and control groups?

Answers to Practice Test 2

1. a 2. a 3. c 4. c 5. b 6. c 7. d
8. $s_p = 6.395$
9. $t = .69$, p-value = .25. There is insignificant evidence that the additive increases gas mileage.
10. $t = 1.58$, p-value = .14. There is an insignificant difference between the repair cost for the two makes of cars..
11. $t = 2.51$, p-value = .013. The reading instruction was beneficial in raising reading scores.
12. $z = 2.60$, p-value = .0097. There is a highly significant difference between the proportion of city and county families that shop at the mall.
13. $t = 1.26$, p-value = .25. There is an insignificant difference between the premiums for the two companies.

14. $t = 2.04$, p-value $= .044$. Significant evidence exist that the new ingredient improves the screening ability of the suntan lotion.

15. $t = .55$, p-value $> .25$. Insignificant evidence that the drug increases maze-learning scores.

16. $t = 1.46$, p-value $= .0744$. There is a slight significant difference between the estimates with one eye closed and both eyes open.

17. $t = -2.75$, p-value $.014$. There is a significant difference in the coordination scores for the experimental and control groups.

10

Analysis of Categorical Data

The objective of this chapter is to present methods of analyzing data that consist of frequency counts for the various categories of one or more categorical variables.

Important issues—Chapter 10

- When data consist of frequency counts and satisfy the properties of a multinomial experiment the statistic given by

$$\chi^2 = \Sigma(o_j - e_j)^2/e_j$$

is used to test hypotheses about the data.

- In this chapter we discussed three applications of this test-- the χ^2 goodness-of-fit test, the χ^2 test of independence and the χ^2 test of homogeneity. To use the statistic in any of the three cases the expected cell counts must be found. For the goodness-of-fit the expected counts are $e_j = n \, \pi_j$ where n is the number of subjects and π_j represents the hypothesized probability of falling in class j.

- For the test of independence and test of homogeneity the expected cell counts are found by $e_j = RC/n$ where R = row total, C = column total, and n = total number of subjects

- The test statistic is approximately distributed as χ^2 with the degrees of freedom being k - 1 in the goodness of fit and (r-1)(c-1) in the test of independence and the test of homogeneity. In all cases, the approximation is valid if no expected cell count is less than 1 and no more than 20% are less than 5.

Glossary

The Multinomial Experiment
- The experiment consists of n identical, independent trials.
- The outcome of each trial falls in one of k categories.

- The probabilities associated with the k outcomes, denoted by $\pi_1, \pi_2, ..., \pi_k$, remain the same from trial to trial. Because there are only k possible outcomes we have $\pi_1 + \pi_2 + ... + \pi_k = 1$.
- The experimenter records the values $o_1, o_2, ..., o_k$, where o_j (j = 1, 2, ..., k) is equal to the observed number of trials in which the outcome is in category j. Note that $n = o_1 + o_2 + ... + o_k$.

Expected Number Of Outcomes In A Multinomial Experiment

Out of n trials of a multinomial experiment the expected number of outcomes to fall in category j is $e_j = n\pi_j$. Remember that the expected cell count is calculated assuming H_0 is true.

Pearson χ^2 Test Statistic

$$\chi^2 = \Sigma(o_j - e_j)^2/e_j$$

where the sum is over all classes with o_j being the observed frequency count and e_j the expected frequency count in class j.

χ^2 Goodness-of-Fit Test

Application: Multinomial Experiments

Assumptions: The experiment satisfies the properties of a multinomial experiment. No expected cell count, e_j, is less than 1 and no more than 20% of the e_js are less than 5.

H_0: $\pi_1 = p_1, \pi_2 = p_2, ... \pi_k = p_k$

where $p_1, p_2, ..., p_k$ are the hypothesized values of the multinomial probabilities.

H_a: At least one of the multinomial probabilities does not equal the hypothesized value.

Test statistic: $\chi^2 = \Sigma(o_j - e_j)^2/e_j$ where $e_j = np_j$

The test is a right-tailed test where the p-value is found in the χ^2 table with k-1 degrees of freedom. Usually the exact value cannot be found, but bounds for it can be found by finding the closest values to the observed value of the χ^2 statistic.

Chi-Square Test Of Independence

Application: Test the independence of two classifying variables.

Assumptions: The experiment satisfies the properties of a multinomial experiment. No expected cell count is less than 1 and no more than 20% are less than 5.

H_0: The two classifications are independent

H_a: The two classifications are dependent

Test Statistic: $\chi^2 = \Sigma(o_j - e_j)^2/e_j$ where o_j represents the observed cell frequencies and e_j represents the expected cell frequencies given by $e_j = RC/n$ where R = row total, C = column total, n = grand total or the total number of subjects

The test is a right-tailed test where the p-value is found in the χ^2 table with (r-1)(c-1) degrees of freedom (r denotes the number of rows and c denotes the number of columns).

Chi-Square Test Of Homogeneity

Application: Contingency table with fixed marginal totals.

Assumptions: A random sample is selected from each of the row category populations. The sample sizes (row marginal totals) are fixed prior to sampling. No expected cell count is less than 1 and no more than 20% are less than 5.

H_0: The populations are homogeneous with respect to the variable of classification.

H_a: The populations are not homogeneous.

Test Statistic: $\chi^2 = \Sigma(o_i - e_i)^2/e_i$ where o_i represents the observed cell frequencies and e_i represents the expected cell frequencies given by $e_i = RC/n$ where R = row total, C = column total, n = grand total or the total number of subjects

The test is a right-tailed test where the p-value is found in the χ^2 table with (r-1)(c-1) degrees of freedom (r denotes the number of rows and c denotes the number of columns).

Practice Test 1

I. Multiple choice

1. The following data represents the observed frequency counts for subjects falling into four different categories.

Category 1	Category 2	Category 3	Category 4
18	24	26	32

For the null hypothesis H_0: $\pi_1 = \pi_2 = \pi_3 = \pi_4$ the expected cell counts are
a) the same as the observed frequency counts.
b) unequal.
c) 25 each.
d) unknown from this information.

2. In Exercise 1 the chi-square goodness of fit should
a) reject H_0 because the p-value > .10
b) reject H_0 because the p-value < .10
c) accept H_0 because the p-value > .10
d) accept H_0 because the p-value < .10

3. For a chi-square statistic of 7.2 with 5 degrees of freedom we have
a) p-value < .01
b) .01 < p-value < .05
c) .05 < p-value < .10
d) p-value > .10

4. To test if a particular sample came from a normal population one could conduct
a) the chi-square goodness of fit
b) the chi-square test of independence
c) the chi-square test of homogeneity
d) the z-test

5. In conducting a chi-square test of independence
a) No expected cell count should be greater than 1 and no more than 20% should be greater than 5.
b) No observed cell count should be greater than 1 and no more than 20% should be greater than 5.
c) No expected cell count should be less than 1 and no more than 20% should be less than 5.
d) No observed cell count should be less than 1 and no more than 20% should be less than 5.

II. Problems

6. Six hundred housewives were asked to compare three brands of soap and select their favorite. Use the following data and determine if they have an equal preference of the three brands.

Brand of soap		
A	B	C
218	220	162

7. Students are given a pretest in algebra and scored as low, medium and high. Their first test in statistics was also scored as low, medium and high. Do the following data indicate that there statistics grade is dependent on their ability in algebra?

	Statistics score		
Algebra score	Low	Medium	High
Low	22	12	3
Medium	15	25	10
High	4	21	16

8. Two hundred males and 250 females were asked which characteristic, appearance or performance, they most preferred in an automobile. From the following data determine if there is a significant difference in the preferences of men and women for the "most liked" automobile characteristic.

	Most liked characteristic	
Sex	Appearance	Performance
Male	75	125
Female	150	100

9. Adult males in 3 age brackets were asked if they were more concerned about their mental health or physical health. Do the following data indicate a relationship between age and concern about mental and physical health?

	Most concerned about	
Age	Mental health	Physical health
30 or less	62	29
between 31 and 45	77	16
between 46 and 65	52	37

10. Past records indicate that a certain department has given 10% A's, 20% B's, 40% C's, 20% D's, and 10% F's. A particular professor gave the following grades.

```
                        Grade
                ---------------------------
                A     B     C     D     F
No. of students 5     10    21    9     15
```

Has the professor deviated significantly from the standard for the department?

11. Psychologists are concerned with a person's choice of a counselor when going for therapy. A random sample of 200 subjects indicated the following choices if they were in need of therapy.

```
Choice of counselor                 Number
----------------------------------------
Psychiatrist                          36
Psychologist                          57
Trained Christian psychologist        33
Minister, priest, rabbi               74
   or spiritual leader
```

Are the different counselors equally preferred?

12. The following data are the results of an attempt to assess the predictive validity of Klopfer's Prognostic Rating Scale (PRS) with subjects who received behavior modification psychotherapy. Scores have been ordered.

```
2.2, 4.1, 4.2, 5.0, 5.4, 5.8, 6.3, 6.6, 6.8, 6.9, 7.1,
7.4, 7.4, 7.7, 8.2, 8.7, 9.4, 9.5, 11.7, 11.9
```

Test to see if this data is coming from a normal population with mean 7 and standard deviation of 3.

13. The state highway patrol desires to investigate the issue of drivers who wear seat belts. A random selection of 400 drivers revealed that 97 were less than 26 years old, 106 were between 26 and 35, 97 were between 36 and 55, and 100 were 56 or older. Of the 97 less than 26, only 23 wear their seat belts on a regular basis, 41 of those in the 26 to 35 age group wear seat belts, 68 of those in the 36 to 55 group wear seat belts, and 26 of those over 55 wear seat belts. Determine if there is a relationship between the age of the driver and whether or not they wear seat belts.

14. An economist is concerned with the opinions of the public on the government's ability to curb inflation. From random samples of four regions of the country she determines the number who feel that the government is doing an adequate job in curbing inflation.

```
                            Opinion
           -----------------------------------------------
Region     Adequate job    Inadequate job    No opinion
East           181              92               27
South          147             138               15
Central        165             103               32
West           126             152               22
```

Do the data indicate that opinions on the government's ability to curb inflation are equally distributed across the four regions of the country?

Answers to Practice Test 1

1. c 2. c 3. d 4. a 5. c
6. $X^2 = 10.84$, df = 2, p-value = .0044. They do not have an equal preference for the three brands.
7. $X^2 = 25.479$, df = 4, p-value < .001. There is evidence that their statistics grade is dependent on their ability in algebra.
8. $X^2 = 22.5$, df = 1, p-value < .001. There is a significant difference between the preferences of men and women for the "most liked" automobile characteristic.
9. $X^2 = 13.069$, df = 2, p-value = 0.002. There appears to be a relationship between age and concern about mental and physical health.
10. $X^2 = 2.53$, df = 4, p-value < .20. It appears that the professor has not deviated significantly from the standard for the department.
11. $X^2 = 22.2$, df = 3, p-value < 0.001. There is highly significant evidence that the different counselors are not equally preferred.
12. $X^2 = 3.19$, p-value > 0.10. There is insufficient evidence to say that the data are not coming from a normal population.
13. $X^2 = 23.852$, df = 3, p-value < 0.001. There is a clear relationship between the ages of drivers and whether or not they wear seat belts.
14. $X^2 = 37.36$, df = 6, p-value < 001. The opinions on the government's ability to curb inflation are not equally distributed across the four regions.

Practice Test 2

I. Multiple Choice

1. A manufacturer wishes to compare three different colors for the label on their product. Following are the color preferences for a random sample of 180 consumers:

Red	Green	Orange
68	62	50

For a chi-square goodness of fit the null hypothesis that the colors are equally preferred is
a) H_0: $\pi_1 = \pi_2 = \pi_3$
b) H_0: $p_1 = p_2 = p_3$
c) H_0: colors are independent
d) H_0: colors are dependent

2. Referring to Exercise 1 the expected cell counts are
a) 68, 62, 50
b) not equal
c) 60, 60, 60
d) 180, 180, 180

3. In Exercise 1 the test statistic is $X^2 = 2.8$. Then
a) p-value > .10
b) .05 < p-value < .10
c) .01 < p-value < .05
d) p-value < .01

4. In conducting a chi-square goodness of fit
a) No expected cell count should be greater than 1 and no more than 20% should be greater than 5.
b) No observed cell count should be greater than 1 and no more than 20% should be greater than 5.
c) No expected cell count should be less than 1 and no more than 20% should be less than 5.
d) No observed cell count should be less than 1 and no more than 20% should be less than 5.

5. The reading ability of fourth grade students in two different school districts were measured. The number of students falling in the three categories were recorded below:

	Reading ability		
	Low	Average	High
District A	21	77	38
District B	34	64	30

For the chi-square test of independence the null hypothesis is
a) H_0: $\pi_1 = \pi_2 = \pi_3$
b) H_0: $p_1 = p_2 = p_3$
c) H_0: School district and reading ability are independent
d) H_0: School district and reading ability are dependent

6. In reference to Exercise 5 the marginal total for District B is
a) 55   b) 136   c) 128   d) 264

7. In reference to Exercise 5 the expected cell count for District B and the Average reading ability cell is
a) 44   b) 128   c) 141   d) 68.4

II. Problems

8. A local radio disk jockey gave away free passes to a rock concert to every 10th caller who called in while her program was on the air. The section of the city where each winner lives was recorded in the following table.

Residence of winner			
Section A	Section B	Section C	Section D
28	13	9	41

Is there evidence that her listeners are equally distributed across the city?

9. A random sample of 100 observations was classified into the following categories:

Category			
A	B	C	D
21	35	19	25

Is there evidence that the categories are not equally likely?

10. Random samples of 100 females and 100 males were classified according to their sex and their choice for one of three meals served at a restaurant.

	Meal		
	A	B	C
Female	16	43	41
Male	24	51	25

Do the data provide evidence that the choices of meals are the same for females and males?

11. A study was conducted to determine if the yield of a particular crop and the type soil it was planted in were dependent. Ninety plots of land were planted with the crop after which the yield was obtained and classified as follows:

		Soil Type	
Yield	Sand	Clay	Loam
Low	18	12	4
Medium	8	10	12
High	4	8	14

Is there evidence that the yield depends on the soil type?

12. The faculty at a university were asked if a new faculty lounge were constructed would they use it. Their responses were categorized as follows:

Opinion on a new faculty lounge

	Would use it on a regular basis	Would use it occasionally	Would not use it
Asst Prof	27	33	17
Assoc Prof	38	24	13
Full Prof	42	29	31

Is there evidence that the opinions are equally shared across the ranks of the professors?

13. College students have insisted on freedom of choice when registering for courses. This semester there were seven sections of an introductory history class. The following table shows the number of students who selected the various sections.

Section	1	2	3	4	5	6	7
No. of students	18	12	25	23	8	19	14

Are the sections equally preferred by the students?

14. Subjects from four different regions of the country were asked if they believe that organized prayer belongs in the public schools. Do the following data show a significant relationship between their opinion on organized prayer and the region of the country in which they live?

	Opinion on organized prayer			
Region	For	Against	No opinion	Total
East	125	79	12	216
South	142	58	9	209
Central	137	67	10	214
West	89	110	24	223
Total	493	314	55	862

15. The coronary blood flow for chronic severe anemia patients was classified as Low, Medium and High. Do the following data indicate that the coronary blood flow of the chronic severe anemia patient differs for males and females?

	Coronary blood flow		
Sex	Low	Medium	High
Males	23	19	8
Females	16	25	9

Answers to Practice Test 2

1. a 2. c 3. a 4. c 5. c 6. c 7. d

8. $X^2 = 28.34$, df = 3, p-value < .001. There is highly significant evidence that here listeners are not equally distributed across the city.

9. $X^2 = 6.08$, df = 3, p-value = .1078. It appears that the categories are equally likely.

10. $X^2 = 6.16$, df = 2, p-value = 0.046. There is significant evidence that the choices are different for females and males.

11. $X^2 = 15.352$, df = 4, p-value = 0.004. Yes, there is highly significant evidence that the yield depends on the soil type.

12. $X^2 = 8.279$, df = 4, p-value = 0.083. There is mildly significant evidence that the opinions are not equally shared across the ranks of the professors.

13. $X^2 = 12.94$, df = 6, p-value = 0.0439. It appears that the sections are equally preferred.

14. $X^2 = 43.005$, df = 6, p-value < .001. Yes, there is a significant relationship between the opinion on organized prayer and the region of the country.

15. $X^2 = 2.133$, df = 2, p-value = .345. There is no significant difference between coronary blood flow for male and female chronic severe anemia patients.

11

Regression Analysis

In this chapter we extend our study of regression analysis by examining *statistical models*. In our study of univariate statistics we viewed the parent population as the population that produces the univariate data in the sample. We now view the *population regression line* as the model that produces the sample data in a scatterplot. We now study the relationship between the response variable y and the predictor variable x via this *linear regression model*.

Important issues—Chapter 11

- Main Tasks for a Regression Analysis
  * Construct a scatterplot of the bivariate data.
  * Propose a statistical model that relates the response and predictor variables.
  * Estimate the parameters in the model.
  * Test the parameters in the model.
  * Check the assumptions of the model.
  * Use the model for predictions and for generally describing the relationship between the two variables.

- Regression is the study of the relationship between a *dependent variable* y and the *independent variables* $x_1, x_2, ---, x_k$. The regression equation is of the form $y = \beta_0 + \beta_1 x_1 + \beta_2 x_2 + --- + \beta_k x_k + \varepsilon$

- If k = 1, the equation is referred to as a *simple linear regression equation*. From the data (x_1, y_1), $(x_2, y_2), ---, (x_n, y_n)$ the *least squares estimates* of β_0 and β_1 are found.

- From the least squares equation the *predicted value of y* is found by the equation $\hat{y} = b_0 + b_1 x_i$ where b_0 and b_1 are the least squares estimates.

- The *residual* associated with the data point (x_i, y_i) is $e_i = y_i - \hat{y}_i$. All the residuals are squared and summed to form SSE, the *sum of squares due to error*.

- The *coefficient of determination* is the square of the correlation coefficient and gives the percent of the variability in the dependent variable that is explained by the independent variable.

- Test of hypothesis about the regression coefficient, β_1, can be performed. The test statistic for testing $H_0: \beta_1 = 0$ is given by

$$t = \frac{b_1}{s/\sqrt{SS_x}}$$

and is distributed as a t-distribution with n-2 degrees of freedom when the basic assumptions are satisfied.

- Confidence intervals for β_1 can be constructed with the formula $b_1 \pm t^* s/\sqrt{SS_x}$.

- Confidence intervals for the expected response, μ_y are found with the formula

$$b_0 + b_1 x' \pm t^* s\sqrt{1/n + (x' - \overline{x})^2/SS_x}.$$

- A prediction interval for a new y is found with the formula

$$b_0 + b_1 x' \pm t^* s\sqrt{1 + 1/n + (x' - \overline{x})^2/SS_x}.$$

- *Multiple regression* is the study of the relationship between the dependent and several independent variables. The computations are best handled with a computer.

- An *analysis of the residuals* makes it possible to check out the assumptions that are made in a regression analysis. It also can point out possible deficiencies in the regression model and suggest alternatives.

Glossary

Linear Regression Model
Given the response variable y and the predictor variable x, the linear regression model that relates the two is $y = \beta_0 + \beta_1 x + \varepsilon$ where β_0 and β_1 are the regression coefficients and ε is the error term.

Assumptions about the error term in the linear regression model $y = \beta_0 + \beta_1 x + \varepsilon$
The error term, ε, is a random variable satisfying the following conditions.
- It has mean value 0 ($\mu_\varepsilon = 0$).
- It has standard deviation σ which does not depend on x.
- It has a normal distribution.
- Any two observed values of ε are independent of each other.

The population regression line is estimated by the least squares regression line given by $\hat{y} = b_0 + b_1 x$

where $b_1 = \dfrac{s_{xy}}{s_x^2} = \dfrac{\Sigma(x_i - \overline{x})(y_i - \overline{y})}{\Sigma(x_i - \overline{x})^2}$ and $b_0 = \overline{y} - b_1 \overline{x}$

Mean Square Error

The mean square error, given by MSE = SSE/(n-2), is an unbiased estimate of σ^2. The square root of MSE is the estimated standard deviation $s = \sqrt{MSE}$.

Assumptions Needed To Make Inferences About The Regression Model

- •y is related to x by the linear regression model $y = \beta_0 + \beta_1 x + \varepsilon$
- •As a random variable, y is normally distributed with a mean of $\mu_y = \beta_0 + \beta_1 x$ and a standard deviation of σ.
- •Realizations of the random variable y are independent.

Test Of Hypothesis About β_1

Left-Tailed Test	Right-Tailed Test	Two-Tailed Test
H_0: $\beta_1 \geq 0$	H_0: $\beta_1 \leq 0$	H_0: $\beta_1 = 0$
H_a: $\beta_1 < 0$	H_a: $\beta_1 > 0$	H_a: $\beta_1 \neq 0$

Standardized Test Statistic:

$$t = \frac{b_1}{SE(b_1)}$$

where $SE(b_1) = s/\sqrt{\Sigma(x_i - \bar{x})^2}$, degrees of freedom = n-2

A $(1-\alpha)100\%$ **confidence interval for** β_1 is given by the limits $b_1 \pm t^* SE(b_1)$

For a given x', a $(1-\alpha)100\%$ **confidence interval for the expected response** μ_y is given by the limits

$b_0 + b_1 x' \pm t^* SE(\hat{y})$ where $SE(\hat{y}) = s\sqrt{1/n + (x' - \bar{x})^2/SS_x}$

For a given x', a $(1-\alpha)100\%$ **prediction interval for an individual y** is given by the limits

$b_0 + b_1 x' \pm t^* s\sqrt{1 + 1/n + (x' - \bar{x})^2/SS_x}$

where t^* is the upper $\alpha/2$ critical value found in the t-table with n-2 degrees of freedom.

Practice Test 1

I. Multiple Choice

1. Regression analysis is a statistical procedure that provides for
a) a comparison of the distributions of a single variable in two populations.
b) a comparison of the distributions of a single variable in several populations.
c) a method of relating two or more variables in a single population.
d) a method of relating only qualitative variables in a single population.

2. The equation $y = \beta_0 + \beta_1 x$ is called
a) a correlation equation.
b) a prediction equation.
c) a simple linear regression equation.
d) a multiple regression equation.

3. A multiple regression equation involves
a) one dependent variable and several independent variables.
b) one independent variable and several dependent variables.
c) several independent and dependent variables.
d) only one independent and one dependent variable.

4. If the null hypothesis that $\beta_1 = 0$ in the linear regression model $y = \beta_0 + \beta_1 x$ is rejected then
a) x and y are not related.
b) x and y are not linearly related.
c) x and y are linearly related.
d) x and y are related but not in a linear fashion.

5. To test the hypothesis that $\beta_1 = 0$ in the linear regression model $y = \beta_0 + \beta_1 x + \varepsilon$ we must assume
a) that the ε's are independent
b) that the ε's are normally distributed
c) that the distribution of the ε's have a common standard deviation
d) all of the above.

6. Suppose $y = 2.8 + 1.5x_1 - .8x_2$. The predicted y for $x_1 = 2$ and $x_2 = 3$ is
a) 8.2
b) 1.6
c) 5.8
d) 3.4

7. In Exercise 6 the residual for the data point (2,3,5) is
a) 8.2
b) 1.6
c) 5.8
d) 3.4

II. Problems

8. Given the following summary data:

$n = 20 \quad \Sigma x_i = 668 \quad \Sigma y_i = 124.6 \quad \Sigma(x_i - \bar{x})^2 = 2974.8$
$\Sigma(y_i - \bar{y})^2 = 1099.64 \quad \Sigma(x_i - \bar{x})(y_i - \bar{y})^2 = -1186.44$

a) Find the least squares regression line.
b) Find a 95% confidence interval for the regression coefficient.

9. Given the following summary data:

$$n = 8 \quad \Sigma x_i = 56 \quad \Sigma y_i = 40 \quad \Sigma(x_i - \bar{x})^2 = 132$$
$$\Sigma(y_i - \bar{y})^2 = 56 \quad \Sigma(x_i - \bar{x})(y_i - \bar{y})^2 = 84$$

a) Find the least squares regression line.
b) Test the significance of the regression coefficient.

10. A retail merchant wishes to determine if a linear relationship exists between the amount spent on advertising and her sales volume. Recorded below are the costs per week on advertising and the weekly sales volume.

x (advertising cost in $100)	7	15	21	20	25	14	9
y (sales volume in $1000)	16	30	32	35	41	33	20

Find the regression line relating advertising cost and sales volume.

11. From the data and the solution found in Exercise 10 find a 99% confidence interval for the regression coefficient.

12. Suppose $y = 5.3 - 2.4x_1 + 1.5x_2$. Find the predicted y for $x_1 = 1$ and $x_2 = 2$.

13. From the following data find the regression equation and test the hypothesis that the regression coefficient is zero, i. e., test H_0: $\beta_1 = 0$.

x	6	9	4	5	7
y	5	3	8	7	4

14. A psychiatrist wanted to know if the level of pathology in psychotic patients 6 months after treatment could be predicted from pretreatment symptom ratings of thinking disturbance. From the following data construct the regression line and test the significance of the regression coefficient.

Patient	1	2	3	4	5	6	7	8
Level of pathology	4	2	1	3	2	7	5	2
Thinking disturbance	2	3	2	4	1	3	3	4

Answers to Practice Test 1

1. c 2. c 3. a 4. c 5. d 6. d 7. b
8. a) $\hat{y} = 19.59 - .4x$ b) $-.4 \pm .227$
9. a) $\hat{y} = .52 + .64x$ b) t= 12.05, p-value < .005
10. The regression equation is $\hat{y} = 9.90 + 1.24 x$
11. $1.24 \pm .86$

12. 5.9

13. $\hat{y} = 11.85-1.04x$, t=-6.39, p-value < .01. The regression coefficient is significantly different from 0.

14. $\hat{y} = 2.33+.333x$, t = .4, p-value > .10. The regression coefficient is insignificant.

Practice Test 2

I. Multiple Choice

1. The equation $y = \beta_0 + \beta_1 x_1 + \beta_2 x_2$ is called
a) a correlation equation.
b) a prediction equation.
c) a simple linear regression equation.
d) a multiple regression equation.

2. If the null hypothesis that $\beta_1 = 0$ in the linear regression model $y = \beta_0 + \beta_1 x$ is not rejected then
a) x and y are not related.
b) x and y are not linearly related.
c) x and y are linearly related.
d) x and y are related but not in a linear fashion.

3. If a simple linear regression model is adequate then the standardized residuals
a) should be zero.
b) should exhibit a linear trend with the dependent variable.
c) should appear to have come from a standard normal population.
d) are of no value.

4. If two variables are negatively correlated then the regression coefficient, β_1 is
a) negative
b) positive
c) zero
d) can't tell from the information

5. A second-degree polynomial model is described with a
a) simple linear regression equation.
b) multiple regression equation.
c) coefficient of determination.
d) none of the above.

6. Suppose $y = 5.6 - 2.4x_1 + 1.6x_2$. The predicted y for $x_1 = 5$ and $x_2 = 3$ is
a) -1.6   b) 1.6   c) 3.4   d) -3.4

7. In Exercise 6 the residual for the data point (5,3,5) is

146

a) -1.6   b) 6.6   c) 3.4   d) -3.4

II. Problems

8. Given the following summary data:

$$n = 12 \quad \Sigma x_i = 147 \quad \Sigma y_i = 131 \quad \Sigma(x_i - \overline{x})^2 = 160.25$$
$$\Sigma(y_i - \overline{y})^2 = 8090.92 \quad \Sigma(x_i - \overline{x})(y_i - \overline{y})^2 = -592.75$$

a) Find the least squares regression line.
b) Find a 98% confidence interval for the regression coefficient.

9. From the following summary data construct a 95% confidence interval for the expected value, E(y), when x = 2.9.

$$n = 25 \quad \Sigma x_i = 76 \quad \Sigma(x_i - \overline{x})^2 = 280.96$$
$$\Sigma y_i = 202 \quad \Sigma(y_i - \overline{y})^2 = 232.84 \quad \Sigma(x_i - \overline{x})(y_i - \overline{y})^2 = -149.08$$

10. Construct the least squares regression line from the following data:

x	-1	-1	0	0	1	1	2	2
y	12	11	8	9	6	5	3	4

11. Suppose y = -12.6 + 4.5x_i - 2.8x_2. Find the predicted y for x_1 = -1 and x_2 = 3.

12. From the following summary data find the regression equation relating x and y.

$$n = 60 \quad \Sigma x_i = 675 \quad \Sigma(x_i - \overline{x})^2 = 5281.25$$
$$\Sigma y_i = 79.8 \quad \Sigma(y_i - \overline{y})^2 = 369.67 \quad \Sigma(x_i - \overline{x})(y_i - \overline{y})^2 = 691.25$$

13. Income per share for a common stock has been recorded for the past 7 years.

Year	1	2	3	4	5	6	7
Income/share	3.0	3.5	3.7	4.0	3.9	4.2	4.7

Find the regression equation relating income per share to the year.

14. Serum cholesterol was measured on 20 men whose ages ranged from 45 to 75.

Age	51	63	48	50	45	53	55	60	46	71
Serum cholesterol	224	247	249	233	218	265	231	244	208	290

Age	73	48	55	52	75	61	49	68	72	47
Serum cholesterol	278	210	238	225	298	275	240	273	288	212

Determine the regression equation relating age and serum cholesterol and find a 90% confidence interval for the regression coefficient.

Answers to Practice Test 2

1. d 2. b 3. c 4. a 5. b 6. a 7. b
8. a) $\hat{y} = 56.24 - 3.7x$ b) (-9.0, 1.6)
9. 8.153 ± 1.073
10. The regression equation is $\hat{y} = 8.60 - 2.70\,x$
11. -25.5
12. $\hat{y} = -.144 + .131x$
13. The regression equation is $\hat{y} = 2.90 + 0.239\,x$
14. $\hat{y} = 102 + 2.54x$, $2.54 \pm .525$

12

Analysis of Variance

Analysis of variance allows one to statistically analyze several population means with a single test. *The one-way ANOVA design* is an experiment in which independent random samples are obtained from the several populations. By comparing the variability between the samples to the variability within the samples with an F-test we are able to determine whether there is a significant difference between the population means.

Important issues—Chapter 12

- The *total variability* in the data for a one-way design is partitioned into two parts, called the *within sample* variability and *between sample* variability. The within sample variability is measured by *the sum of squares for error* and has $n_1 + n_2 + \text{---} + n_k$ - k degrees of freedom where, n_j is the sample size for the sample from the jth population. The between sample variability is measured by the *sum of squares for treatment* and has k - 1 degrees of freedom. Dividing the degrees of freedom into the sum of squares yields a *mean square*. Dividing the mean square for treatments by the mean square for error produces the *F-statistic* which is used to test the hypothesis of equality of the population means.

- In order to conduct the F-test, certain assumptions about the populations should be satisfied. Those assumptions are that the populations should be normally distributed and have homogeneous variances. If the assumptions are not met, then the *Kruskal-Wallis* test should be used.

- Checking Assumptions
  * Examine dotplots and boxplots for unusual behavior.
  * Check the normality assumption be construction normal probability plots for each of the k samples.
  * Check the homogeneity of variances assumption by comparing the largest sample variance to the smallest sample variance. If the largest s^2 is no more than 4 times as large as the smallest s^2 assume that the assumption is satisfied. Equivalently the largest standard deviation, s, should be no more than twice as large as the smallest standard deviation.

- If an analysis of variance indicates a significant difference in the population means, then *Tukey's multiple comparison procedure* is used to further analyze the means.

- If you are using a calculator the following computational formulas for the sum of squares will be useful. (Refer to the notation in Table 12.1)

Total sum of squares = TSS = $\Sigma\Sigma y_{i,j}^2 - T^2/N$

Sum of squares for treatments = SST = $\Sigma(T_j^2/n_j) - T^2/N$

Sum of squares for error = SSE = TSS - SST

The value $C = T^2/N$ appears frequently and is called the correction term.

Glossary

The **sum of squares for error** is $SSE = (n_1-1)s_1^2 + (n_2-1)s_2^2 + \cdots + (n_k-1)s_k^2$

$$= \Sigma(y_{i,1} - \overline{y}_1)^2 + \Sigma(y_{i,2} - \overline{y}_2)^2 + \cdots + \Sigma(y_{i,k} - \overline{y}_k)^2$$

and has $n_1 + n_2 + \cdots + n_k - k$ degrees of freedom. Dividing the degrees of freedom into the sum of squares for error yields the **mean square for error**

$$MSE = \frac{SSE}{df} = \frac{(n_1-1)s_1^2 + (n_2-1)s_2^2 + \cdots + (n_k-1)s_k^2}{n_1 + n_2 + \cdots + n_k - k} = s_p^2$$

The **sum of squares for treatments** is $SST = \Sigma n_j(\overline{y}_j - \overline{y}_G)^2$ and has $k-1$ degrees of freedom. Dividing the degrees of freedom into the sum of squares yields the **mean square for treatments**

$$MST = \frac{SST}{df} = \frac{\Sigma n_j(\overline{y}_j - \overline{y}_G)^2}{k-1} = s_b^2$$

Analysis Of Variance For The One-Way ANOVA
 Assumptions:
- The samples are randomly and independently selected from their respective populations.
- The sampled populations are normally distributed.
- The variances of the sampled populations are equal; that is, $\sigma_1^2 = \sigma_2^2 = \cdots = \sigma_k^2$.

H_0: $\mu_1 = \mu_2 = \cdots = \mu_k$ versus H_a: at least two μ's differ.

Test Statistic: $F = MST/MSE$

The test is a right-tailed test where the p-value is found in the F-table with $k-1$ and $n_1 + n_2 + \cdots + n_k - k$ degrees of freedom.

The **total sum of squares** (SSTotal) in a data array consisting of k samples is the total of the squared deviations between each observation and the overall mean. If $y_{i,j}$ denotes the ith observation in the jth sample and $\bar{y}_G$ denotes the overall mean then SSTotal $= \Sigma\Sigma(y_{i,j} - \bar{y}_G)^2$ and has $n_1 + n_2 + \text{---} + n_k - 1$ degrees of freedom. An important relationship between SSTotal, SST, and SSE is SSTotal = SST + SSE.

A **(1-α)100% confidence interval** for the mean of treatment j is given by the limits $\bar{y}_j \pm t^* \sqrt{MSE/n_j}$ where t^* is the upper $\alpha/2$ critical value found in the t-table with N-k degrees of freedom.

Tukey's Multiple Comparison Procedure

For a specified value of α, calculate $W = q_\alpha(k,v)\sqrt{MSE/n}$ where

n = number of observations in each sample
MSE = the mean square error from the ANOVA table.
k = number of different population means
v = degrees of freedom associated with MSE
$q_\alpha(k,v)$ = critical value of the Studentized range found in Table B7 in Appendix B.

To conduct Tukey's procedure complete the following steps.

- Rank the sample means from highest to lowest and order the population means in the same order.
- Compute the difference between the largest and smallest sample means: $\bar{y}_{largest} - \bar{y}_{smallest}$. If the difference exceeds W, then the corresponding population means are declared significantly different. Proceed to compute the difference between the largest and the next smallest sample mean: $\bar{y}_{largest} - \bar{y}_{2ndsmallest}$ As above, if the difference exceeds, W then declare the corresponding population means different. Continue to make comparisons with the largest sample mean, $\bar{y}_{largest} - \bar{y}_{3rd\ smallest}$ and so on, until a difference fails to exceed W. Once a difference between two sample means is less than W, the corresponding population means, and all means between, are declared nonsignificant.
- Next, make comparisons with the next largest sample mean, $\bar{y}_{2nd\ largest} - \bar{y}_{smallest}$ and so on, using the same procedures as in step 2. Continue until all possible comparisons are made.
- Summarize the results by drawing a line under the population means that are declared nonsignificant.
- The procedure for the Kruskal-Wallis test, which is presented here, is the ordinary F-test applied to the rank summary data, n_j, $\bar{y}_{Rj}$, and s_{Rj}. This is an approximation of the original Kruskal-Wallis test, and therefore it is suggested that there are *five or more observations in each sample*. For samples less than 5, exact tables for the Kruskal- Wallis test can be found in Conover (1980).

Practice Test 1

I. Multiple choice

1. $F_{.01,3,20} =$
a) 2.38 b) 5.18 c) 26.7 d) 4.94

2. Suppose an F-test in an ANOVA involving 3 treatment levels with 12 observations in each sample gave an F value of 6.61. Then the
a) p-value > .10
b) .05 < p-value < .10
c) .01 < p-value < .025
d) p-value < .01

3. Consider the following partially completed ANOVA table.

```
     SV          SS        df
-----------|--------|------
Treatment     286.3       4
Error         821.4      60
```

If there are an equal number of observations in each group then there were how many subjects in each group?
a) 12   b) 13   c) 15   d) 65

4. In Exercise 3 the F-statistic is
a) 5.23   b) 13.69   c) 2.87
d) not enough information to determine its value

5. Below are the rank orderings of three California burgundies by four wine experts.

Tester	Brand A	Ranks Brand B	Brand C
A	2	1	3
B	2	1	3
C	1	2	3
D	1	2	3

These data should be analyzed by
a) the ordinary F-test.
b) Tukey's multiple comparison test.
c) the Kruskal-Wallis test.
d) none of the above because the samples are not independent.

6. Assumptions for the Kruskal-Wallis test include
a) the parent distributions are normal in shape.
b) the sample sizes all exceed 30.
c) the parent distributions are similar in shape.
d) all of the above.

II. Problems

7. From the following summary data complete an ANOVA table for an F-test. Include the p-value.

n	10	10	10
$\bar{y}$	72.8	65.3	74.1
s	10.6	9.2	13.7

8. Following are elasticity readings for random samples drawn from three different processes for making plastic.

Process A	Process B	Process C
4.1	5.8	3.6
2.6	6.3	2.2
3.7	5.9	3.6
2.7	4.7	3.2
2.1	5.4	1.5
3.5	7.1	2.7
3.3	6.6	3.8
3.9	7.3	3.0

Perform an analysis of variance to determine if there is a difference in the mean elasticity readings for the three processes.

9. Perform Tukey's multiple comparison using $\alpha = .05$ on the following data.

Group	1	2	3	4	5
n	10	10	10	10	10
$\bar{y}$	3.2	2.8	4.8	5.0	3.7
s^2	.46	.51	.39	.20	.28

10. Below are the rank orderings of three independent samples. Is there a significant difference in the rankings of the three groups?

Ranks		
Group A	Group B	Group C
2	1	3
4	7	6
5	8	11
10	9	12

11. A home heating contractor sells 3 types of oil heaters. To compare the heating units, the following efficiency ratings were obtained on samples of each type heater:

Type A	Type B	Type C
75	73	60
71	83	63
74	70	74
86	66	56
77	54	61
84	71	73
76	74	71
75	76	62
57	92	82
96	75	64

Do the data provide sufficient evidence of a difference in the mean efficiency ratings of the 3 types of heating units?

12. A sociologist is interested in whether or not occupational level of parents affects the anxiety scores of 9th grade students. The following data represents scores on the anxiety test for pupils from homes of parents in four different occupational levels.

Occupational Level			
1	2	3	4
8	23	21	14
6	11	21	11
12	17	22	12
16	16	18	10
10	14	9	13
21	14	6	11
12	17	22	12
16	16	18	10
10	14	9	13

a) Construct an ANOVA table and test the appropriate hypothesis.
b) Analyze the means using Tukey's Multiple comparison test

13. Following are cortisol levels for 3 groups of pregnant women who delivered between 38 and 42 weeks gestation. Group1 was elective Caesarean section, Group2 is emergency Caesarean, and Group3 delivered vaginally. Is there a significant difference in cortisol level for the 3 groups?

```
Group1   262  307  211  323  454  339  304  154  287  355

Group2   465  501  455  355  468  362

Group3   343  772  207 1048  838  687
```

14. A pharmaceutical company has three suppliers of a particular raw material for their product. In a effort to compare the quality of the materials, the levels of impurities were measured in sample batches from the three suppliers.

Supplier A	1.52	1.61	1.84	1.56	1.47	1.60	1.54	1.59	1.64	1.26
Supplier B	1.57	1.39	1.41	1.46	1.38	1.42	1.76	1.44	1.47	1.20
Supplier C	1.33	1.47	1.48	1.40	1.12	1.66	1.42	1.71	1.49	1.24

Do the data provide sufficient evidence of a difference in the mean impurity levels from the 3 suppliers?

15. The debate concerning where introductory statistics should be taught has been going on for a long time. Some feel that each department should teach their own introductory statistics course. Others feel that the introductory course should be taught in a single statistics department. At one university introductory statistics is taught in the College of Business, the College of Education, and the Statistics department in the College of Arts and Sciences . Students of psychology, sociology, political science, and criminal justice take their statistics in the Statistics department. After completing their courses students in each of the three colleges were given a comprehensive test on basic statistical concepts. The test was approved by the instructors in the three areas. Do the following data indicate that students in any one area perform better at the introductory level of statistics?

	Business	Education	Statistics
2		4	
3	1	6	
4	2	0 3 6	
5	0 5	0 4 5 7	0 5
6	0 0 1 2 3 5 6	0 2 4 7 8 9	0 4 5 7
7	0 0 2 3 4 5 5 8 9 9	0 0 2 3 5 7	0 0 1 2 3 4 5 5 7 9
8	0 1 2	0 2 7	0 4 5 5 9
9	2	1	0 5 9
10			0

Answers to Practice Test 1

1. d 2. d 3. b 4. a 5. d 6. c

7.

SV	SS	df	MS	F	p-value
Treatment	451.267	2	225.6335	1.76	> .10
Error	3462.21	60	128.23		

8.

SV	SS	df	MS	F	p-value
Treatment	49.741	2	24.870	39.82	< .01
Error	13.118	21	0.625		

9. $w = .775$ μ_4 μ_3 μ_5 μ_1 μ_2

$$\underline{\hspace{3em}}\quad\underline{\hspace{3em}}$$
$$\underline{\hspace{3em}}$$

10.

SV	SS	df	MS	F	p-value
Treatment	15.5	2	7.8	0.55	>.10
Error	127.5	9	14.2		

11.

SV	SS	df	MS	F	p-value
Treatment	567.3	2	283.6	3.16	.05 -.10
Error	2421.7	27	89.7		

12.

SV	SS	df	MS	F	p-value
Treatment	142.3	3	47.4	2.52	.076
Error	602.7	32	18.8		

$w = 5.56$ μ_3 μ_2 μ_1 μ_4

$$\underline{\hspace{6em}}$$

13.   Analysis of Variance

Source	DF	SS	MS	F	p
Factor	2	458296	229148	7.52	0.004
Error	19	578671	30456		
Total	21	1036967			

14. $F = 2.07$, p-value = .145

15. $F = 5.46$, p-value = .006

155

Practice Test 2

I. Multiple Choice

1. $F_{.05,4,15} =$
a) 2.36
b) 3.87
c) 3.06
d) 5.86

2. The assumptions for an analysis of variance include
a) the sampled populations are normally distributed.
b) the sampled populations have the same variance.
c) the samples are independently selected from the populations.
d) all of the above.

3. Analysis of variance gets it name from the fact that
a) the test is for comparing population variances.
b) the test is for comparing population means by investigating two sources of variability.
c) both of the above.
d) neither of the above.

4. Suppose an F-test in an ANOVA involving 4 treatment levels with 15 observations in each sample gave an F value of 4.72. Then the
a) p-value > .10
b) .05 < p-value < .10
c) .01 < p-value < .05
d) p-value < .01

5. Suppose three samples are selected of sizes 18, 20, 24. Their standard deviations were ,respectively 12.4, 15.8 and 14.6. The pooled standard deviation is
a) 14.267
b) 14.336
c) 14.468
d) 14.415

6. Consider the following partially completed ANOVA table.

```
       sv          ss        df
 -----------|--------|------
 Treatment     128.4        3
 Error         352.8       28
```

The F statistic is
a) 2.75
b) 3.4
c) 42.8
d) 12.6

7. Three judges were asked to rank seven contestants on the uneven parallel bars.

	Rank		
Contestant	Judge 1	Judge 2	Judge 3
1	3	2	2
2	7	5	6
3	1	1	3
4	4	3	5
5	6	7	7
6	2	4	1
7	5	6	4

These data should be analyzed by
a) the ordinary F-test.
b) Tukey's multiple comparison test.
c) the Kruskal-Wallis test.
d) none of the above because the samples are not independent.

II. Problems

8. Three methods of instruction in group-encounter techniques were to be compared with respect to the mean level of group interaction. A total of 15 group leaders participated in the study; 5 were assigned technique A, 4 to technique B, and 6 to technique C. After one week of training all leaders were assigned to an encounter group and were scored on their ability to achieve meaningful group interaction.

	Score	
Technique A	Technique B	Technique C
82	72	90
80	30	92
81	79	83
83	75	89
84		87
		93

Do the data indicate that the three instructional techniques have

9. The following ranks were collectively assigned to three independent samples. Complete an ANOVA table on the ranks for the Kruskal-Wallis test. Include the p-value.

Sample 1	Sample 2	Sample 3
4	2	1
6	5	3
10	9	7
11	12	8
15	13	14
19	18	16
20	22	17
21	25	23
27	26	24

10. In an auto emission study the reduction in oxides of nitrogen was measured using three different types of additives to the gasoline. From the following data determine if there is a significant difference in the mean reduction in oxides of nitrogen for the three additives.

Additive	Reduction in oxides of nitrogen						
A	22	14	38	15	26	19	20
B	36	38	29	26	34	30	41
C	15	19	24	13	28	16	14

11. From the following summary data complete an ANOVA table for an F-test. Include the p-value.

n	15	15	15	15
$\bar{y}$	6.2	8.1	4.3	9.4
s	1.82	2.34	1.65	1.96

12. Independent random samples were selected from three populations with the following results:

```
-- |---------------------|---------------------|----------------
 4 |                     | 6                   |
-- |---------------------|---------------------|----------------
 5 | 5                   | 3 6 7               | 3 5
-- |---------------------|---------------------|----------------
 6 | 2 4                 | 4 5 7               | 5 7 8
-- |---------------------|---------------------|----------------
 7 | 2 3 5 7 8           | 0 1 2 2 4 5 5 7 9   | 2 3 4 6 7 9
-- |---------------------|---------------------|----------------
 8 | 0 1 2 2 3 4 6 8 9 9 | 0 2 5 6 9           | 0 1 2 3 5 6 7 9
-- |---------------------|---------------------|----------------
 9 | 2 3 4 6 8           | 2 4 8               | 3 4 7 9
-- |---------------------|---------------------|----------------
10 | 0 0                 | 0                   | 0 0
```

a) Complete side-by-side box plots and comment on the shapes of the distributions.
b) To compare the distributions, should an ordinary F-test or the Kruskal-Wallis test be conducted?

13. Using the data in Exercise 12, test the hypothesis that the population means are equal using the F-test suggested in part (b) of Exercise 12.

14. To determine if the testing environment has any effect on test scores three groups of students were given the Verbal Scholastic Aptitude Test (SATV) in different environments. Group A took the test in their homerooms in the presence of their regular teacher. Group B also took the exam in their homerooms, but were proctored by strangers. Group C took the exam in an unfamiliar setting, proctored by a stranger. Their SAT scores were as follows:

Group A	Group B	Group C
370	410	380
390	460	390
420	350	310
560	430	270
480	420	530
440	280	480
470	520	360
740	450	620
430	420	480
290	640	430
470	410	410
690	340	370

Do the data suggest that the testing environment has an effect on the SAT scores?

15. The following data is the results of a comparison of four different treatments of a particular disease and the resulting recovery rates.

Treatment	No of days to recovery									
A	3	7	5	9	8	4	9	6	4	8
B	7	8	9	5	6	12	9	10	8	5
C	4	3	6	2	5	3	7	4	3	6
D	7	8	5	9	7	6	5	11	9	10

Determine if there is a significant difference in the mean recovery rates for the four treatments.

Answers to Practice Test 2

1. c 2. d 3. b 4. d 5. d 6. b 7. c
8. $F = 5.59$, p-value = .019
9. **Analysis of Variance**

Source	DF	SS	MS	F	p
Factor	2	28.2	14.1	0.21	0.812
Error	24	1609.8	67.1		
Total	26	1638.0			

10. $F = 10.19$, p-value = .001
11.

SV	SS	df	MS	F	p-value
Treatment	223.5	3	74.5	19.41	< .01
Error	214.9294	56	3.838		

12.

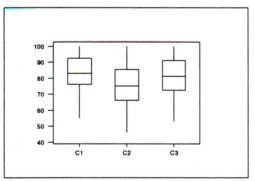

   a) All three distributions appear reasonably symmetrical.

   b) The ordinary F-test can be applied to these data.

13. **Analysis of Variance**

Source	DF	SS	MS	F	p
Factor	2	793	397	2.37	0.101
Error	72	12075	168		
Total	74	12868			

159

14. **Analysis of Variance**

Source	DF	SS	MS	F	p
Factor	2	25356	12678	1.11	0.341
Error	33	376408	11406		
Total	35	401764			

15. **Analysis of Variance**

Source	DF	SS	MS	F	p
Factor	3	82.70	27.57	6.56	0.001
Error	36	151.20	4.20		
Total	39	233.90			

Individual 95% CIs For Mean

```
                                        Based on Pooled StDev
 Level    N      Mean     StDev   ------+---------+---------+---------+
 A        10     6.300    2.214                 (------*-----)
 B        10     7.900    2.234                         (-----*------)
 C        10     4.300    1.636    (------*-----)
 D        10     7.700    2.058                     (-----*------)
                                   ------+---------+---------+---------+
 Pooled StDev =     2.049               4.0       6.0       8.0      10.0
```

The Statistics Tutor

Part C

Solutions to Odd-Numbered Exercises

Your solutions and the solutions given in this manual may not agree completely because of round-off error. Most solutions were found with the Minitab computer package.

1

Collecting And Understanding Data

Section 1.1 Essential Elements of Statistics

1.1   Experimental unit: College student
      Population: All college students at this university
      The 500 students should be selected in some random manner such as selecting every 10th student
      from a list provided by the registrar.

1.3   Experimental unit: An accident
      The population is not clearly identified because the exercise doesn't specify what region the
      accidents were taken from. For example, did all accidents occur in a particular city or state?
      Variables: Date accident occurred, How many vehicles involved, Make and model of vehicles,
      How many occupants, Gender of driver, Age of driver, Was alcohol involved?
      The sample size is 1,000.

1.5   Population: All entering freshmen at the university
      Gender, SAT, 1st choice, 2nd choice, residence
      From a list provided by the admissions office, sequentially number all freshmen from 1 to n (the
      last number on the list) and then with a computer randomly generate 200 numbers between 1 and
      n and then choose those students.

1.7   a   It would be impossible to test the drug on **all** potential users of the product.
      b   If we distributed free samples to everyone then no one would be required to purchase the
          product.
      c   It would be impossible to count every tree in the forest.
      d   If we determined the life of every battery we would deplete our inventory.

1.9   a   13.67 per 100, 7.67 per 100, 6.67 per 100, 5.11 per 100, 5.00 per 100.

      b   **Character Dotplot**

```
      ..              .        .                                           .
     +---------+---------+---------+---------+---------+-------rate
     4.8       6.4       8.0       9.6      11.2      12.8
```

c   No, just because 7.6% is the average for these five stores, this doesn't mean that every store will have a 7.6% error rate.   Some stores will possibly have a 0% error rate but then others will go much higher than the average.

d   No, in fact the exercise stated that 9 stores were inspected and only 5 were fined for overcharges.

1.11   The target population consist of potential customers for the new shopping center.
The sampled population consist of people appearing on the chamber of commerce mailing list. Taking every 10th name is okay as long as the list is random with respect to the questions addressed in the questionnaire.

1.13   The target population consist of older women. Variables of interest include age, weight, and bone density.

1.15   **Character Dotplot**

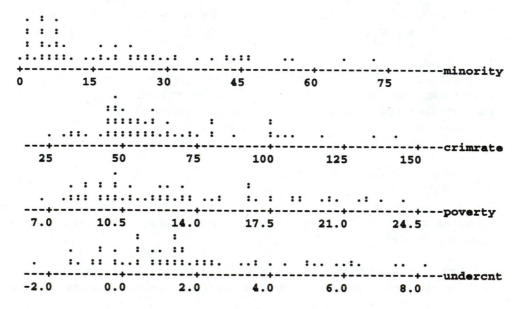

The distributions for crimerate, poverty, and undercount seem to be similar in shape. All four distributions have observations that extend out on the right tail.

Section 1.2 Sources of Data

1.17   The most general population for which this sample is representative is the population of people whose characteristics are similar to those who subscribe to *Time* magazine.

1.19   No, a large segment of rock music fans never attend concerts.

1.21   First, assign everyone in the class a number from 1 to n (the size of the class). Then generate 15 random numbers from 1 to n and select those students.

1.31   The *Digest* prediction (the worst ever made in a national poll) was inaccurate for several reasons. First, it was a voluntary survey; only 23% of those receiving cards returned them. Generally voluntary surveys are biased because only those who feel strongly about the issue will bother to respond. Second, the sample of names was taken from subscription lists of magazines, telephone directories and automobile registrations. This was clearly biased towards those with higher incomes and better education. This is especially important because this poll was made during the depression era. Simply stated, the sample was not representative of the population. It had an overrepresentation of the upper middle class and the more highly educated. There were many laborers and farm workers at that time, and it is likely that they were not properly represented in the sample. (George Gallup observed the bias in the *Digest*'s poll and predicted that Roosevelt would win the election with 56% of the popular vote. This was the beginning of the nationally recognized Gallup Poll.)

Section 1.3   Critically Appraising Data

1.33   It is only possible to determine the percent of the police force that is **caught** taking a bribe. It would be impossible to determine the percent that has never taken a bribe.

1.35   During this time frame it became more acceptable to report sexual abuse of children. It may not mean that there was in increase in the number of cases, but rather there was an increase in the number of **reported** cases. The size of the children in the rectangles is misleading because both the height and width are increased. In examples such as this, only one dimension should be increased.

1.37   This is a captive audience, not a random sample of disaffected Clinton supporters. They were presented only one side of the story, namely Clinton's speech. Furthermore, it doesn't say how the 100 people were selected. This information should not be generalized to the population of all disaffected Clinton supporters.

1.39   If a school has a large number of transfer students then their graduation rate will be reduced because all those who transfer are counted as not graduating. Unless there are no transfers, the actual graduation rates would be higher than those given here. To avoid the problem the NCAA could completely eliminate transfer students from the calculations.

1.41   Although the Russians were given the most tickets, they were not the worst offenders because their rate was 8.9 tickets/vehicle/month and Ukraine diplomats had a rate of 10.6 tickets/vehicle/month.

Section 1.4 Surveys, Experiments, Observational Studies

1.43   No. The neighbor who was at home was there for a given reason; perhaps he/she was unemployed or had several small children or any number of things that would characterize them as being different from the person who was chosen for the survey. The interviewer should return at another time to obtain responses from the selected household; otherwise, another household should be randomly selected.

1.45   A strong point is that educational information concerning nuclear reactors as a source of energy is provided. A weak point is that the results may be biased because of the pamphlet. Because the respondents choose to participate (volunteer survey) it is questionable whether or not the results will represent the true feeling of the population.

1.47   A variable is confounded with the treatment when its effect and the effect of the treatment cannot be distinguished from each other.

1.49   The placebo effect is a psychological effect to a treatment.

1.51   This question leads the respondent by stating, "Since over half of all fatal traffic accidents involve alcohol----". If the "over half" phrase is used it should be used in a more neutral manner such as,

"Over half of all fatal traffic accidents are alcohol related. Do you agree or disagree that the penalties for drunk driving should be increased?"

1.53   This question leads the respondents.   In option a, the phrase "to those countries that protect the rights of their citizens" should be omitted.

1.55   a   Because the TV viewing is controlled, this is a randomized experiment.
         b   The treatment is type of TV program viewed by the children. The response variable is the number of violent acts by the children.
         c   A potentially confounding variable is the initial hostility level. The random assignment to the two groups, however, should "average out" any initial differences.

1.57   a   observational study
         b   randomized experiment
         c   observational study
         d   randomized experiment

1.59   a   sample survey
         b   experiment
         c   sample survey

1.61   This is not a scientific poll, it is a volunteer survey. Viewers of the Weather Channel choose to participate by volunteering to respond to the survey. They certainly would not have a random sample of Weather Channel viewers.

1.63 a   The population of interest is the potential shoppers in the 200 stores around the country.
      b   Because there are only 200 stores we could use random digits to select a simple random sample of the stores and then randomly select shoppers in the selected stores.
      c   The target population is the potential shoppers in the 200 stores and the sampled population is the current shoppers in the 200 stores. In this case the target and sampled populations are not exactly the same but it is likely that the characteristics of the two are very similar.
      d   Variables of interest include the age, gender, annual income, etc of the shopper.

1.65 a   This is an observational study.
      b   The treatment is employment status. The response variable is the divorce rate.
      c   Extraneous variables that would affect divorce rate are years married, stress, and life style.

1.67 This is convenience sampling because no scientific method is used to select the respondents.

1.69 a   telephone
      b   mailed questionnaire
      c   mailed questionnaire
      d   personal interview

1.71 Systematic sampling would be the easiest and because there is no order other than being listed alphabetically, the sample would be representative of all charge accounts.

Section 1.5 Summary and Review

1.73 a   Population of interest: All oil stocks
      b   Sample: 10 selected oil stocks
      c   Variable: Price/earning ratio

1.75 Personal income.

1.77 Lower income households are omitted from the sample. Higher income households are less likely to participate.

1.79 D   1.81   A   1.83   D

1.85 treatment   1.87   confounded

1.89 a   T
      b   F
      c   F
      d   F
      e   F

167

1.91 a   Amount of exercise
      b   Risk of heart disease
      c   A bus driver or policeperson in New York City
      d   Job stress and marital status
      e   Observational study

1.93 The phrase "Don't you agree…" is leading. A better question would be 'Do you agree or
      disagree that a farmer should be allowed to raise as much of any crop he/she chooses without any
      restrictions or supports from the federal government?"

1.95 a   Cluster sampling
      b   Stratified sampling
      c   Systematic sampling
      d   Lottery sampling
      e   Stratified sampling

1.97 No, because this is an observational study, we cannot conclude causation. There are too many
      possible confounding variables. Having a higher income and being married probably contributed
      to the decision to own a pet, not the other way around.

1.99 Sample survey

1.103 **Character Dotplot**

```
       . .
      : : :
      : : : .
      : : : : :   .    .
      +---------+---------+---------+---------+---------+-------indians
      .
      :
      : .
      : : : : .        .   . : .     : .  : .  . .                  .          redsox
      +---------+---------+---------+---------+---------+-------redsox
      :
      :
      :
      : :
      : :
      : : . . . :    .       . .    .    .    .    .       .          angels
      +---------+---------+---------+---------+---------+-------angels
      :
      : . .
      : : : :  . : . . . . . .  .    . .    .    .              orioles
      +---------+---------+---------+---------+---------+-------orioles
      0       1000      2000      3000      4000      5000
```

All distributions have salaries that extend out on the right tail, some more than others. The
salaries for the indians are much less spread out than the other distributions. They have no
players with very high salaries. The redsoxs pay their players considerable more than the other
teams.

1.105

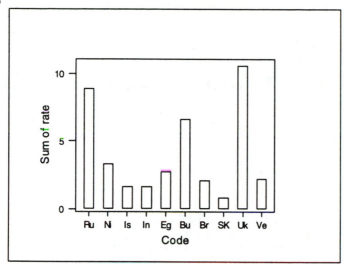

From the chart we easily see that the diplomats from the Ukraine are the worst offenders. We also see that the diplomats from S. Korea are the least offenders.

2

Organizing And Summarizing Univariate Data

Section 2.1 Types of Data

2.1   a   Population of interest is all welfare recipients in the state.
        b   A systematic sample from a list of welfare recipients.
        c   Gender, race, and marital status of head-of-household
        d   Annual income (continuous), number of children (discrete), and age of head-of-household (continuous)

2.3   a   Population is all third graders.
        b   Use a cluster sample of third grade classes.
        c   Gender, race, and availability of a computer at home.
        d   Family income (continuous), IQ (continuous), and number of siblings (discrete)

2.5   a   Numerical-- continuous
        b   Categorical
        c   Numerical-- continuous
        d   Numerical-- continuous
        e   Numerical-- continuous
        f   Numerical-- discrete
        g   Categorical

2.7   Gender—categorical, Major—categorical, Grade point average—numerical, Number of times using placement service—numerical, Type of employment—categorical.

2.9   a   Threshold reaction time is numerical.
        b   It is continuous.

2.11   A categorical variable measures some attribute of an experimental unit which classifies the unit into a certain non-numeric category. A numerical variable measures some numeric characteristic of each experimental unit.

2.13   a   Continuous
        b   Continuous
        c   Discrete
        d   Continuous

e   Discrete
f   Continuous

2.15   a   An experimental unit is a state.
       b   This is a multivariate data set.
       c   There are four variables.
       d   Region—categorical, Hazardous waste sites—numerical, Above average minority—
           numerical, Percent minority—numerical.
       e   Hazardous sites—discrete, Above average minority—discrete, Percent minority—
           continuous

2.17

2.19

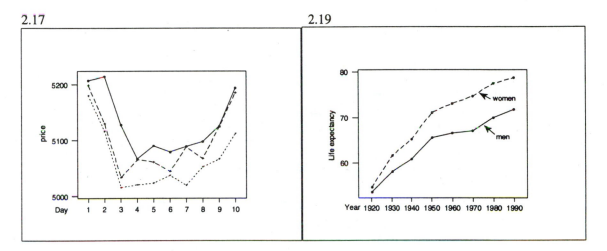

2.21

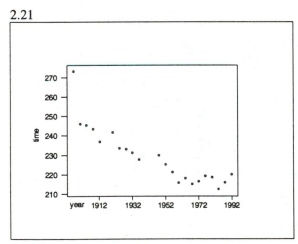

       a   For those years without data leave a space for the year and don't plot a dot for its time.
       b   There was a sudden drop from 1896 to 1900. Over the years there has been a steady
           downward trend. There appears to be an upward trend for the last three Olympic years.
       c   The data are not cyclical. There is a downward trend.

Section 2.2 Displaying Categorical Data

2.23

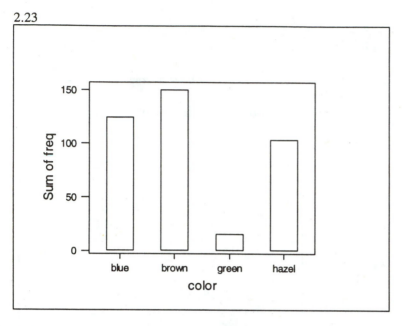

2.25

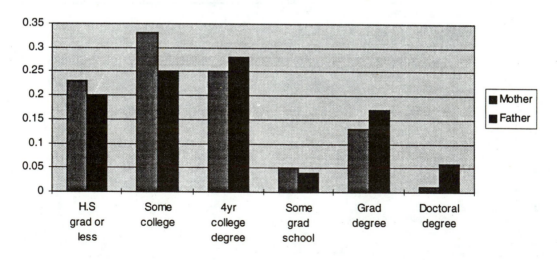

Educatonal Background

2.27 a

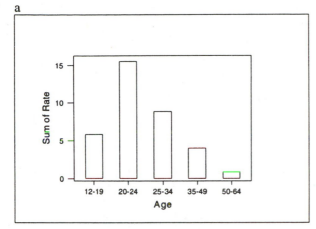

b   We know that a total of 572,032 women victims of domestic abuse and we know the rate
     per 1,000 women for each age group. We do not know, however, how many women are in
     each age group. We would need that information to find the number of women who are
     victims in each age group.

2.29

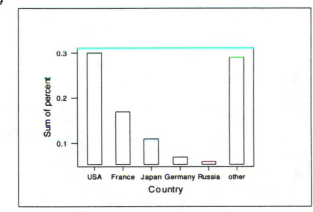

The percents do not add to 100%, therefore we add a category called other.

2.31	Crime	number	rel freq
	Murder	4	0.000735
	Forcible rape	89	0.016360
	Robbery	136	0.025000
	Aggravated assault	630	0.115809
	Burglary	1439	0.264522
	Larceny theft	2946	0.541544
	Vehicle theft	166	0.030515
	Arson	30	0.005515

a   Larceny theft had the highest relative frequency of 54.1544%
b   Larceny theft and Vehicle theft accounts for 57.2% of all crimes.

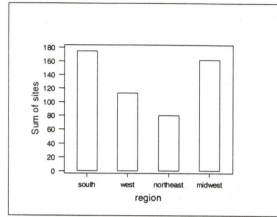

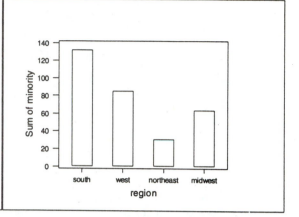

2.35

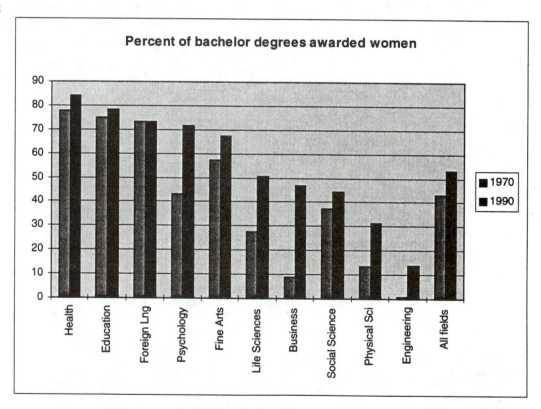

Percent of bachelor degrees awarded women

b   Psychology, Life Sciences, Business, Physical Sci, and Engineering show a dramatic change from 1970.

c   Comparison pie charts would not show the difference as well as the comparison bar graph.

d   The graphical bar graph shows the differences much better than does the table.

Section 2.2 Displaying Categorical Data

2.37   `Stem-and-leaf of miller    N = 20`
        `Leaf Unit = 1.0`

```
        2     1 67
        3     1 8
        5     2 01
       10     2 22233
       10     2 5555
        6     2 67
        4     2 9
        3     3 01
        1     3 3
```

2.39   `Stem-and-leaf of number    N = 20`
        `Leaf Unit = 0.10`

```
        3     6 000
        7     7 0000
       (6)    8 000000
        7     9 0000
        3    10 000
```

2.41   `Stem-and-leaf of score    N = 23`
        `Leaf Unit = 1.0`

```
        2     0 67
        5     0 889
        8     1 001
       (4)    1 2223
       11     1 4444555
        4     1 67
        2     1 89
```

2.43   `Stem-and-leaf of loss    N = 30`
        `Leaf Unit = 1.0`

```
        2    -1 20
        4    -0 65
        8    -0 4432
       14     0 223344
      (12)    0 555556778899
        4     1 001
        1     1 5
```

2.45   **Stem-and-leaf of weight   N = 25**
Leaf Unit = 1.0

```
    1     11  7
    2     12  9
    4     13  26
    6     14  48
   12     15  014479
  (4)     16  0226
    9     17  3579
    5     18  34
    3     19  07
    1     20  9
```

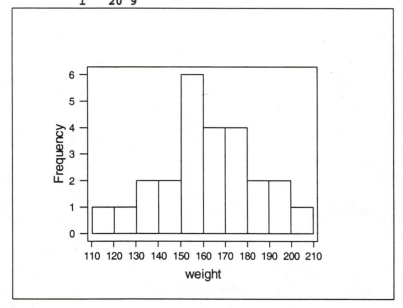

2.47

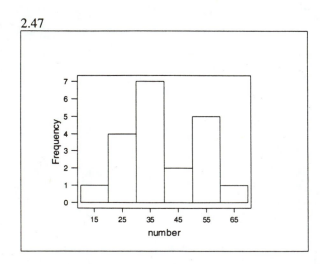

2.49

```
Stem-and-leaf of score      N = 24
Leaf Unit = 1.0

     1     4 3
     2     4 8
     2     5
     5     5 589
     8     6 234
    12     6 6689
    12     7 34
    10     7 5579
     6     8 1234
     2     8 6
     1     9 1
```

2.51   2.53

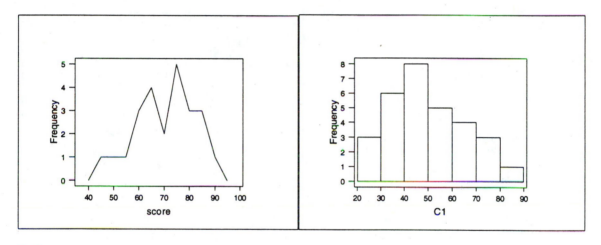

2.55 a   Stem-and-leaf of cholest   N = 62
          Leaf Unit = 10

```
     1     1 67
     4     1 84,92,98
    13     2 00,02,10,11,12,15,16,17,18
    30     2 20,25,25,26,30,30,30,30,31,32,32,32,34,34,36,36,38
   (11)    2 40,43,46,47,48,54,54,54,56,56,58
    21     2 63,64,67,67,67,68,70,70,72,78,78
    10     2 83,85
     8     3 00,00,08
     5     3 27,34,36
     2     3 53
     1     3
     1     3 93
```

b

Class limits	Class boundaries	Frequency
160 - 179	159.5 - 179.5	1
180 - 199	179.5 - 199.5	3
200 - 219	199.5 - 219.5	9
220 - 239	219.5 - 239.5	17
240 - 259	239.5 - 259.5	11
260 - 279	259.5 - 279.5	11
280 - 299	279.5 - 299.5	2
300 - 319	299.5 - 319.5	3
320 - 339	319.5 - 339.5	3
340 - 359	339.5 - 359.5	1
360 - 379	359.5 - 379.5	0
380 - 399	379.5 - 399.5	1

c   d

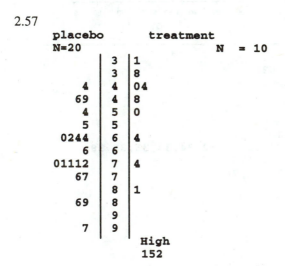

2.57

```
    placebo        treatment
    N=20                        N   =  10
              3 |1
              3 |8
         4    4 |04
        69    4 |8
         4    5 |0
         5    5 |
      0244    6 |4
         6    6 |
     01112    7 |4
        67    7 |
              8 |1
        69    8 |
              9 |
         7    9 |
                 High
                 152
```

178

Section 2.4 Summary Measures of Location

2.59 a $\bar{y} = 24.4$ $\bar{y}_{T.10} = 24.25$

b $\bar{y} = 26.4$ $\bar{y}_{T.10} = 24.25$ The mean increased by 2 points and the trimmed mean remained the same.

c   If the 16 is changed to 10 the mean will decrease but the trimmed mean will not be affected. This illustrates that the mean is affected by observations on the tails of a distribution but the trimmed mean is resistant to extreme scores on the tails of a distribution.

2.61 a $\bar{y} = 33.36$ $\bar{y}_{T.10} = 34.23$

b $\bar{y} = 23.36$ $\bar{y}_{T.10} = 24.23$ Both the mean and trimmed mean are reduced by 10 points.

c   If a constant is add (or subtracted) from all scores then the mean and trimmed mean will increase (or decrease) by an amount equal to the constant.

2.63 a $\bar{y} = 24.64$ $\bar{y}_{T.10} = 25.187$

b $\bar{y} = 2.464$ $\bar{y}_{T.10} = 2.5187$ The mean and trimmed mean are also divided by 10.

c   If each score is multiplied or divided by a constant, then the mean and trimmed mean are also multiplied or divided by the constant.

2.65

Low	Q_1	M	Q_3	High
48	93	98.5	106	149

2.67 The mean and median are both equal to 41 mpg. Because the data are very symmetrical a trimmed mean will be very much the same.

2.69 $\bar{y} = 176{,}024$ $M = 163{,}586$

Character Dotplot

```
    :. : .:.        .      ..  ... ..     .              .          .
  -+---------+---------+---------+---------+---------+---------+-----populat
  70000     140000    210000    280000    350000    420000
```

The two extreme observations on the right tail will increase the value of the mean but will have no effect on the median. Because they are on the right tail (positive direction) the mean will be larger than the median.

2.73 **Descriptive Statistics**

Variable	N	Mean	Median	TrMean	StDev	SEMean
sites	51	10.39	8.00	8.96	11.14	1.56
minority	51	6.08	3.00	4.84	8.65	1.21

Variable	Min	Max	Q1	Q3
sites	0.00	54.00	2.00	16.00
minority	0.00	53.00	0.00	10.00

179

```
Letter Value Display

        DEPTH       LOWER        UPPER         MID         SPREAD
N=      51
M       26.0                8.000                          8.000
H       13.5        2.000       16.000          9.000       14.000
E        7.0        1.000       19.000         10.000       18.000
D        4.0        0.000       28.000         14.000       28.000
C        2.5        0.000       36.500         18.250       36.500
B        1.5        0.000       48.500         24.250       48.500
         1          0.000       54.000         27.000       54.000


Letter Value Display

        DEPTH       LOWER        UPPER         MID         SPREAD
N=      51
M       26.0                3.000                          3.000
H       13.5        0.500       10.000          5.250        9.500
E        7.0        0.000       13.000          6.500       13.000
D        4.0        0.000       14.000          7.000       14.000
C        2.5        0.000       19.500          9.750       19.500
B        1.5        0.000       38.500         19.250       38.500
         1          0.000       53.000         26.500       53.000
```

2.75 $\bar{y} = 113.45$ $M = 119$ $\bar{y}_{T.10} = 115.12$

Section 2.5 Summary Measures of Variability

2.79 a   Deviations from the mean are -0.7, -1.8, 1.7, 0.9, and -0.1. Clearly, they sum to 0.
      b   SS $= (-0.7)^2 + (-1.8)^2 + (1.7)^2 + (0.9)^2 + (-0.1)^2 = 7.44$
      c $s = 1.364$

2.81

```
Stem-and-leaf of math        N  = 30
Leaf Unit = 1.0

      1      3 8
      6      4 02234
      9      4 599
     (7)     5 0123444
     14      5 5789
     10      6 01234
      5      6 57
      3      7 01
      1      7 5


      ȳ = 54.9,  s = 9.75
```

The data are close to being bell shaped so we apply the Empirical rule. Approximately 68% of the data should be within one standard deviation of the mean, that is, between the limits 45.15 to 64.65. The actual percent is 18/30 = 60%. Approximately 95% of the data should be within two standard deviations of the mean, that is, between the limits 35.4 and 74.4. The actual percent is 29/30 = 96.67%. All the data are within three standard deviations of the mean.

2.83 $\bar{y}_1 = 78.58$, $s_1 = 6.73$ $\bar{y}_2 = 80.81$, $s_2 = 9.06$

A score of 82 in class one is $z = (82 - 78.58)/6.73 = .51$ standard deviations above the mean. A score of 82 in class two is $z = (82 - 80.81)/9.06 = .13$ standard deviations above the mean. The score of 82 is higher in class one.

2.85   **Letter Value Display**

	DEPTH	LOWER	UPPER	MID	SPREAD
N=	28				
M	14.5	148.000		148.000	
H	7.5	142.000	160.500	151.250	18.500
E	4.0	142.000	167.000	154.500	25.000
D	2.5	141.000	174.000	157.500	33.000
C	1.5	139.500	183.000	161.250	43.500
	1	138.000	190.000	164.000	52.000

Low	Q1	M	Q3	High
138	142	148	160.5	190

Q-spread = 18.5   Range = 52

The large difference between the Q-spread and the range is caused by the long right tail of the distribution. The Q-spread is not affected by observations on the tails of the distribution, whereas, the range is completely determined by the two end observations.

2.87

Low	Q1	M	Q3	High
618	1478	2129	2774	3298
	MidQ=	2126		
	MidR=	1958		

The median and the midQ are very close to each other because the middle 50% of these data are symmetrical causing the average of Q1 and Q3 to be very close to the middle (median).

2.89

Low	Q1	M	Q3	High
14	20	23	26	42
	MidQ=	23		
	MidR=	28		

```
Stem-and-leaf of age        N  = 87
Leaf Unit = 1.0

     4      1  4555
    10      1  667777
    19      1  888889999
    35      2  0000001111111111
   (14)     2  22222223333333
    38      2  4444444445555
    25      2  66667
    20      2  888889
    14      3  0
    13      3  2333
     9      3  5
     8      3  666
     5      3  8899
     1      4
     1      4  2
```

The median and the midQ are the same because the middle 50% of the data are symmetrical. When this happens the average of Q1 and Q3 will be very close to the middle (median). The midRange is so much larger than the median because of the data on the right tail of the distribution. The high score of 42 increases the value of the midRange.

2.93   Because the data have a very long tail on the right the midRange will be pulled toward that tail causing it to be much larger than the median. For the same reason the Range will be much larger than either the Q-spread or E-spread.

2.95   a $\overline{y} = 113.45$, $s = 35.79$

        b   Between 77.66 and 149.24 (one standard deviation of the mean) we find 11/20 = 55%, and between 41.87 and 185.03 (two standard deviations of the mean) we find 19/20 = 95% of the data. According to the Empirical rule we should find approximately 68% of the data within one standard deviation where we found 55%. This probably due to the fact that the data are not bell shaped. We get perfect agreement, however, within two standard deviations (95%). A stem and leaf plot would reveal that the data are not bell shaped.

        c   The z-score for 165 is $z = (165 - 113.45)/35.79 = 1.44$; that is, 165 is 1.44 standard deviations above the mean.

Section 2.6 Describing the Shape of a Distribution

2.97   The midsummaries become progressively smaller as we scan down from the median to the midRange. This is supporting evidence that the distribution is skewed left.

2.99   The midsummaries are all about the same, suggesting that the distribution is symmetrical. We cannot deduce that the distribution has long tails based on the midsummaries. We need the boxplot to see the outliers.

2.101 These midsummaries have no distinct pattern. This is supporting evidence that the distribution is bimodal.

2.103 The midsummaries get progressively small as we scan down from the median until we get to the midRange. The one observation, 128, causes the midRange to go back up. In summary we should classify the shape of this distribution as skewed left.

2.105 The test scores in class2 are more variable than those in class1. Class2 has higher scores but the median score in class1 is higher than the median of class2.

2.107 Based on the appearance of the boxplot we should classify this distribution as skewed right (positively skewed). The outlier and long tail on the high side will inflate the value of the mean. They will have no effect on the median worth. The median is more representative of a typical score.

2.109 a   **Stem-and-leaf of 1993**   **N = 51**
Leaf Unit = 100

```
      6     0  011111
     17     0  22222233333
    (11)    0  44455555555
     23     0  666666677777
     11     0  889999
      5     1  011
      2     1  2
      1     1     There is one unusual observation, 2832.8, for D.C.
      1     1
      1     1
      1     2
      1     2
      1     2
      1     2
      1     2  8
```

b   **Letter Value Display**

	DEPTH	LOWER	UPPER	MID	SPREAD
N=	51				
M	26.0	535.500		535.500	
H	13.5	328.750	758.150	543.450	429.400
E	7.0	211.500	977.300	594.400	765.800
D	4.0	130.900	1119.700	625.300	988.800
C	2.5	117.600	1164.650	641.125	1047.050
B	1.5	96.400	2020.000	1058.200	1923.600
	1	83.300	2832.800	1458.050	2749.500

The range is much larger than the Q-spread because of the one unusually large observation.

c   **Character Boxplot**

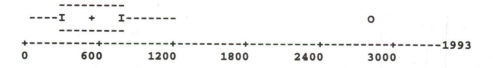

183

The one large observation shows up as an extreme outlier on the boxplot

d   Without the one extreme observation the distribution appears to be reasonably symmetrical. It would be advisable to analyze these data without the one extreme observation.

2.111 a   This distribution is bimodal.

       b   Because this university has an engineering school, it would stand to reason that the two peaks are caused by the salaries of the engineering professors and the salaries of all other professors.

       c   Because these data are made up of two distinct groups it would be inadvisable to average all the scores to get a measure of a typical salary.

Section 2.7 Summary and Review

2.113

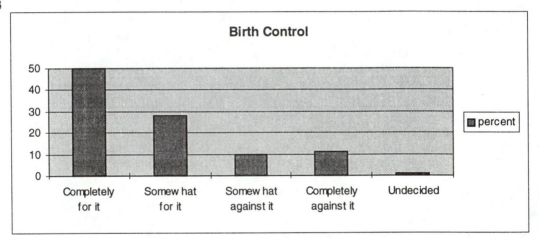

2.115 a   Numerical, Discrete   b   Numerical, Continuous

       c   Categorical   d   Numerical, Continuous

       e   Numerical, Continuous

2.117 a   Categorical   b   Numerical   c   Numerical

       d   Categorical   e   Categorical   f   Numerical

2.119   **Stem-and-leaf of Age   N = 25**
        Leaf Unit = 1.0

```
    2     2 68
    2     3
    5     3 568
    7     4 22
   11     4 5558
   (5)    5 00144
    9     5 799
    6     6 012344
```

2.121

Crime	number	rel freq
murders	198	0.012487
forcible rape	627	0.039543
robberies	5824	0.367306
aggravated assaults	9207	0.580663
Total	15856	

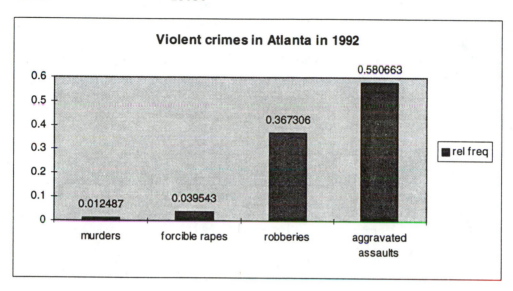

2.123 a   Gender, Age, Marital status, Characteristics of heart received, Length of surgery.

     b   Categorical, Numerical, Categorical, Categorical, Numerical.

2.125 parameter

2.127 a   F

     b   F

     c   T

     d   T

     e   F

     f   F

2.129 a $\overline{y} = 10$, $M = 12$

     b $\overline{y}_{T.10} = 10.714$

     c $s = 4.09$

2.131 a   Population of interest is all registered voters in your district.

     b   A systematic sample from a list of registered voters.

     c   Gender, race, and marital status

     d   Annual income (continuous), number of children (discrete), and age (continuous)

2.133 Corporate name—categorical, Market area—categorical, Assets—numerical, Market percent—numerical.

2.135   a   Numerical   b   Numerical   c   Categorical
       d   Numerical   e   Categorical

2.137   a   Continuous   b   Continuous   c   Discrete
       d   Continuous   e   Continuous

2.139

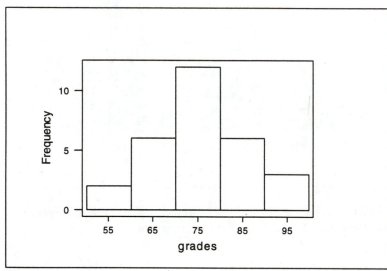

The histogram is bell shaped.

2.141   a   Categorical   b   Numerical   c   Categorical
       d   Categorical   e   Numerical

2.143   a

```
Stem-and-leaf of score      N  = 16
Leaf Unit = 1.0

    3      4 589
    8      5 01444
    8      5 89
    6      6 2
    5      6 68
    3      7 033
```

b	DEPTH	LOWER	UPPER	MID	SPREAD
N=	16				
M	8.5	56.000		56.000	
H	4.5	50.500	67.000	58.750	16.500
	1	45.000	73.000	59.000	28.000

c　The midsummaries increase slightly, indicating a slight skewness to the right.

d　**Character Boxplot**

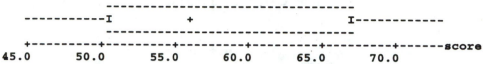

The boxplot also shows a slight skewness to the right. The midsummary analysis and the boxplot, however, overlook an important aspect of the data. The stem and leaf plot shows that the data are bimodal.

2.145

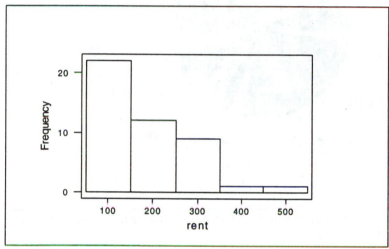

a　The histogram is skewed right.

b　**Descriptive Statistics**

Variable	N	Mean	Median	TrMean	StDev	SEMean
rent	45	179.4	150.0	171.6	101.4	15.1

Variable	Min	Max	Q1	Q3
rent	60.0	490.0	95.0	242.5

The mean and median are very different. Three standard deviations from the mean is
$179.4 + 3(101.4) = 483.6$. There is only one observation, 490, more than 3 standard deviations from the mean.

c　**Letter Value Display**

	DEPTH	LOWER	UPPER	MID	SPREAD
N=	45				
M	23.0	150.000		150.000	
H	12.0	95.000	235.000	165.000	140.000
E	6.5	80.000	302.500	191.250	222.500
D	3.5	72.500	340.000	206.250	267.500
C	2.0	65.000	425.000	245.000	360.000
	1	60.000	490.000	275.000	430.000

The midsummaries definitely increase as we go down from the median to the midRange. The distribution very definitely is skewed right.

187

2.147

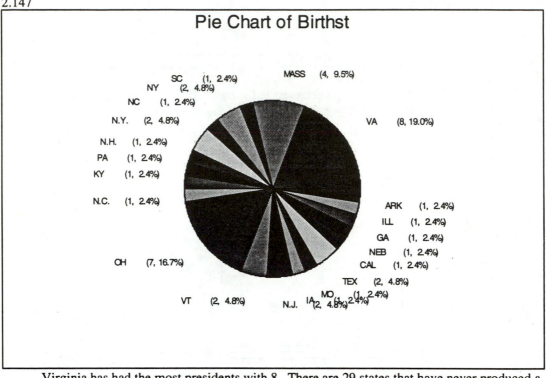

Pie Chart of Birthst

SC (1, 2.4%)
NY (2, 4.8%)
NC (1, 2.4%)
N.Y. (2, 4.8%)
N.H. (1, 2.4%)
PA (1, 2.4%)
KY (1, 2.4%)
N.C. (1, 2.4%)
OH (7, 16.7%)
VT (2, 4.8%)
N.J. (2, 4.8%)
IA (1, 2.4%)
MO (1, 2.4%)
TEX (2, 4.8%)
CAL (1, 2.4%)
NEB (1, 2.4%)
GA (1, 2.4%)
ILL (1, 2.4%)
ARK (1, 2.4%)
VA (8, 19.0%)
MASS (4, 9.5%)

Virginia has had the most presidents with 8. There are 29 states that have never produced a president.

b   **Stem-and-leaf of Inaugage   N = 42**
Leaf Unit = 1.0

```
    2      4 23
    2      4
    5      4 667
    8      4 899
   14      5 001111
   16      5 22
   (8)     5 44445555
   18      5 6667777
   11      5 8
   10      6 0111
    6      6 2
    5      6 445
    2      6
    2      6 89
```

```
Stem-and-leaf of Deathage   N  = 37
Leaf Unit = 1.0

    2      4 69
    3      5 3
    7      5 6778
   13      6 003344
   (6)     6 567778
   18      7 0111234
   11      7 7889
    7      8 013
    4      8 58
    2      9 00
```

Both distributions seem to be bell shaped.

c   **Character Dotplot**

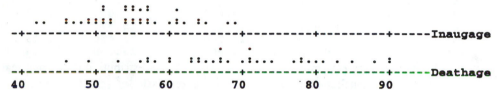

From the dotplots we see that Deathage is more variable than Inaugage and is centered about 15 years above the inaugural age.

2.149 a   From the following stem and leaf plot we see one unusually low score of 4.5 for Alaska and one unusually large score of 22.8 for Florida

```
Stem-and-leaf of percent    N  = 51
Leaf Unit = 1.0

    1      0 4
    1      0
    1      0
    3      1 11
   12      1 222333333
   20      1 44455555
   (20)    1 66666666666677777777
   11      1 8888888999
    1      2
    1      2 2
```

b   **Character Dotplot**

```
                                    .
                            :    . ::.:..
                         ...: .::....:.::::::::.::....
           .                                              .
    +---------+---------+---------+---------+---------+-------percent
   3.5       7.0      10.5      14.0      17.5      21.0
```

Without the largest and smallest scores the distribution is somewhat symmetrical.

c   **Descriptive Statistics**

Variable	N	Mean	Median	TrMean	StDev	SEMean
percent	51	15.965	16.300	16.093	2.807	0.393

Variable	Min	Max	Q1	Q3
percent	4.500	22.800	14.100	17.600

There is not much difference between the mean, median, and trimmed mean. Because of the two outliers on the two tails the trimmed mean is a better representation of the center of the distribution.

d   **Letter Value Display**

	DEPTH	LOWER	UPPER	MID	SPREAD
N=	51				
M	26.0		16.300	16.300	
H	13.5	14.150	17.600	15.875	3.450
E	7.0	13.200	18.600	15.900	5.400
D	4.0	12.300	19.400	15.850	7.100
C	2.5	11.700	19.650	15.675	7.950
B	1.5	8.000	21.300	14.650	13.300
	1	4.500	22.800	13.650	18.300

For the most part the midsummaries indicate that the distribution is symmetrical. The midRange is smaller because of the low score for Alaska.

2.151 $\bar{y} = 1.91$, M $= 1.90$, $\bar{y}_{T.10} = 1.87$, s $= 0.307$

2.153 a   **Stem-and-leaf of pounds   N = 29**
    Leaf Unit = 0.10

```
    4      0 2348
   11      1 0233467
   (7)     2 0224469
   11      3 34459
    6      4 333
    3      5 46
    1      6
    1      7
    1      8
    1      9 1
```

One plant has an unusually large toxic intensity.

b $\bar{y} = 2.757$, M $= 2.430$, s $= 1.913$

There is a large difference between the mean and median because of the one unusually large observation. The median is a better representation of the toxic intensity of these plants.

c   According to the Empirical rule the limits for the middle 95% of the data are -1.069 and 6.583. Only the one observation (9.12) falls outside these limits. Because this one observation inflates the mean and the standard deviation, it is questionable whether we should be applying the Empirical rule.

2.155 From the following time series plot we see a very steady upward trend for the winning average miles per hour for the Indianapolis 500 for the first 14 years. For the last 20 years, however, there doesn't seem to be any pattern. In recent years there has been a lot of variability in the winning speeds. No generalizations can be made over the entire 34 year period.

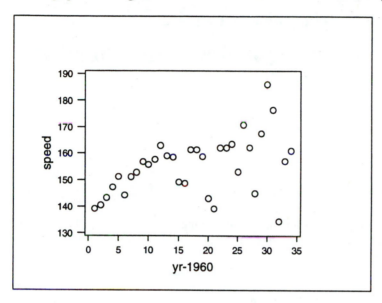

3

Bivariate Data—
Studying Relations Between Variables

Section 3.1 Contingency Tables

3.1   a   50/228 = 21.9%
        b   2/64 = 3.125%
        c   A convicted black defendant, when the victim is white, is 7 times more likely to be given
            the death penalty than a convicted white defendant, when the victim is black.

3.3

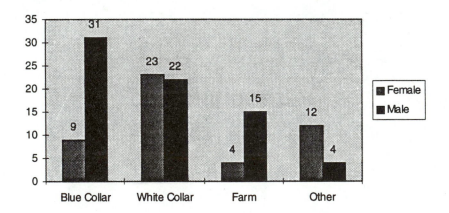

Occupation group by gender

3.5   a and b

	upper	lower	Total
women	20	10	30
men	30	40	70
Total	50	50	100

    c   10/30 = 33.3%
    d   40/70 = 57.1%
    e   10/50 = 20%
    f   men

g   lower

3.7 a   34/57 = 59.65%
     b   23/57 = 40.35%
     c   34/61 = 55.74%
     d   No, in part a the percent is relative to the dealerships that replace parts unnecessarily and in part c the percent is relative to the foreign dealerships.

3.9

	Students	Faculty	Administration	Total
Favor	140	40	25	205
Oppose	360	60	5	425
Total	500	100	30	630

     a   425/630 = 67.5%
     b   360/500 = 72%
     c   60/630 = 9.5%
     d   The figures in the graph reflect frequencies and not relative frequencies.
     e

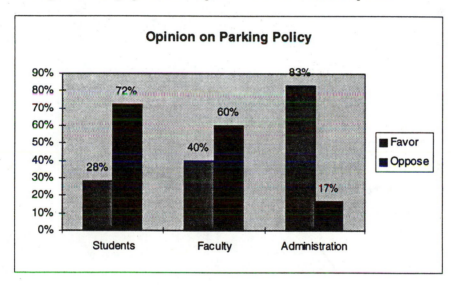

3.11 a   74/104 = 71.15%
     b   74/314 = 23.57%
     c   The percentages in parts a and b are different because they are computed relative to different marginal totals. In part a the percent is relative to all histological type LP. In part b the percent is relative to all positive responses.
     d   126/538 = 23.42%

Section 3.2 Scatterplots

3.13 a

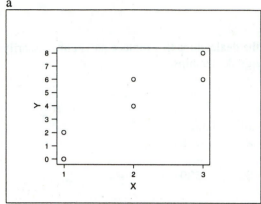

There appears to be a straight line relationship between X and Y.

b

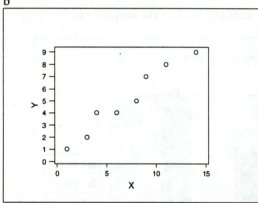

There appears to be a straight line relationship between X and Y.

c

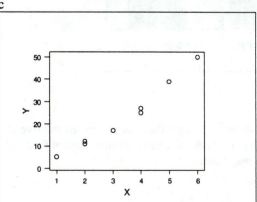

There is a relationship between X and Y which is close to a straight line but also shows a little curvature.

d

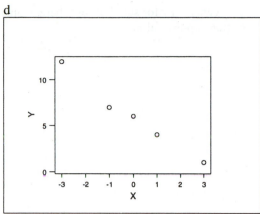

There appears to be a straight line relationship between X and Y with a downward trend.

3.15

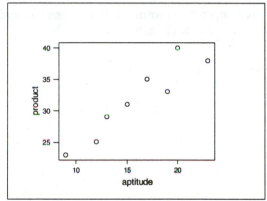

The relationship is a straight line upward (positive) trend. As aptitude increases so does productivity.

3.17

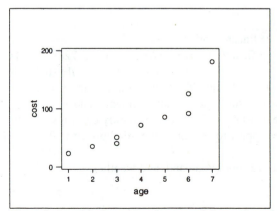

The relationship between the age of the cash register and the maintenance cost for the first 5 years is almost a perfect straight line. The observation at year 7 may suggest a sharp upturn in maintenance cost for older machines; however, we cannot say for sure because there is only one observation. If that trend continues then the relationship is nonlinear.

3.19

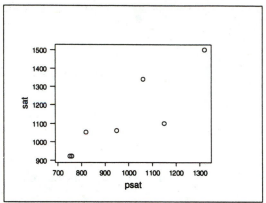

There is one bivariate outlier. Otherwise there is an upward (positive) straight line relationship. Those students with high PSAT scores tend to have high SAT scores.

3.21 a

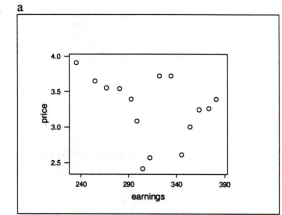

The trend prior to 1987 was a negative linear trend.

b   In 1988 the price of wheat jumped significantly from $2.57/bushel to $3.72/bushel and remained at $3.72/bushel in 1989. In 1990 the price plummeted to $2.61/bushel. There may have been government supports in 1988 and 1989 to explain the high prices.

c   Since 1990 the price has increased with a substantial positive linear trend.

d   There is no natural trend in the price of wheat as it relates to weekly earnings over the observed period of years. There does appear to be a negative trend prior to 1987 and a positive trend after 1990.

e   Two variables that affect wheat price are supply/demand, and weather.

3.23   a

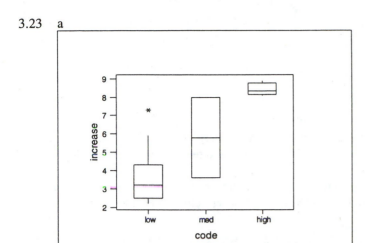

b   When the inflation rate is low and medium there is a lot of variability in the pay increase. There is very little variability in pay increase when the inflation rate is high.

3.25   a

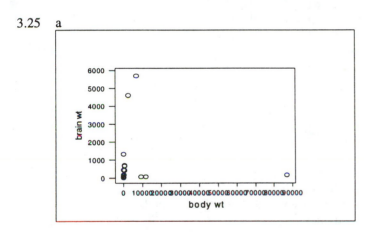

b

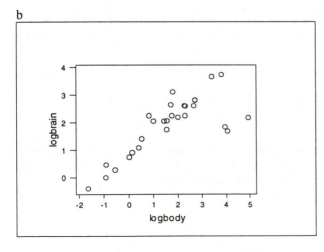

Section 3.3 Correlation

3.27 a r = 0.911
      b r = 0.977
      c r = 0.985
      d r = -0.991

The only correlation that may seem inconsistent with the scatterplot is part a. At each x there is a a degree of variability in the two observed y values that would tend to lower the correlation.

3.29 Because there are no outliers and the relationship appears linear, there is little need to calculate Spearman's correlation. As expected, $r_s = -0.841$, which is very close to Pearson's correlation.

3.31 a

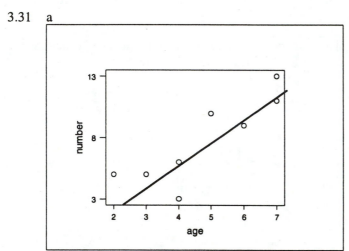

      b r = 0.859 This is consistent with the scatterplot.
      c The relationship is positive. Normally we would expect that as a child's age increases so does the number of gymnastic activities.

3.33 There is very little difference between Spearman's and Peason's correlations (r = -0.884 r_s = -0.885) because there are no outliers and the relationship is close to a straight line.

3.35 a Because one variable is already ranked we should rank the other variable and calculate Spearman's correlation.
      b Pearson's correlation is not appropriate because one variable has already been ranked.
      c $r_s = -0.967$

3.37 Correlation of value and revenue = 0.940. Correlation of value and revenue without outlier = 0.876. The correlation is greater with the outlier than without because it is in the same straight line plane as the rest of the data. Instead of the outlier having an adverse effect on the straight line relationship it actually improves the relationship.

3.39 The correlation of increase and inflatio = 0.815 suggest that a straight line should fit the data reasonably well, the scatterplot, however, suggest otherwise. As indicated in the solution of exercise 3.22, there is a sharp increase in the pay increase when the inflation rate jumps from 5% to 7% but after 7% the pay increase is very steady even for inflation rates up to 14%. We should not use a straight line to describe the relationship over the entire period.

3.41 Correlation of 1994 and 1995 = 0.822. Based on the correlation and the scatterplot, a straight line will fit the data reasonably well.

Section 3.4 Least Squares Regression

3.43
a Response variable is high blood pressure. Predictor variable is air pollution.
b Response variable is mental retardation. Predictor variable is lead poisoning.
c Response variable is death rate for automobile accidents. Predictor variable is percent of drivers wearing seatbelts.
d Response variable is suicide rate. Predictor variable is alcohol consumption among teenagers.
e Response variable is public education expenditures. Predictor variable is per capita income.

3.45
a The regression equation is $Y = -1.67 + 3.00 X$
b The regression equation is $Y = 0.545 + 0.636 X$
c The regression equation is $Y = -6.85 + 8.92 X$
d The regression equation is $Y = 6.00 - 1.80 X$

3.47
a -0.33333, -1.33333, -1.33333, 1.66667, 0.66667, 0.66667
b -0.181818, -0.454545, 0.909091, -0.363636, -0.636364, 0.727273, 0.454545, -0.454545
c 2.93082, 0.01258, -3.82390, 1.25786, -1.82390, 1.01258, 3.33962, -2.90566
d 0.6, -0.8, 0.0, -0.2, 0.4

3.49 The residual plot has a very definite pattern which suggest that the straight line is not the best relationship between the two variables. As pointed out in the solution to exercise 3.14 a straight line will fit the data when the house size is less than 2000 sq ft and another straight line will fit the remaining data for houses whose size exceeds 2000 sq ft.

3.51 $r^2 = 87.5\%$, that is, 87.5% of the variability in productivity is explained by the aptitude scores. The straight line is a reasonable relationship between aptitude and productivity.

3.53
a The regression equation is cost = -16.1 + 23.0 age
b 16.1216 and -12.8919

3.55
a The regression equation is life = 1136 - 3.22 heat
b -100.143, 69.714, 108.571, -68.571, 33.286, -42.857
c SSE = 34323

3.57   a   r = 0.822

        b   The regression equation is y = 2.2 + 0.910x where x is 1994 ratings and y is 1995 ratings.

        c r^2 = 67.6%

        d $(0.822)^2$ = 0.675684 which when rounded is .676 = 67.6%.

        e   A residual plot would help.

3.59   Correlation of Newfound and GrandBk = 0.973

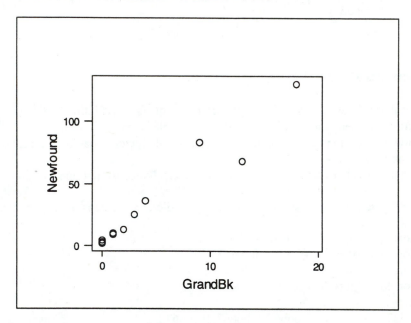

The correlation between the number of icebergs sighted south of Newfoundland and south of the Grand Banks in 1920 is very high (0.973). The scatterplot also shows this strong linear relationship. Although it is not clear which variable (if either) should be the response variable and which should be the predictor variable, it is apparent that the number of sightings at one location is strongly linearly related to the number of sightings at the other location.

Section 3.5 The Resistant Line

3.61 x_L= 4.1, y_L = -3.6
        x_M = 7.65, y_M = 3.95
        x_R = 12.7, y_R = 8.75
        $\hat{y}_R$ = -8.67 + 1.44x

3.63 $x_L = 145, \ y_L = 27$
       $x_M = 180, \ y_M = 18$
       $x_R = 215, \ y_R = 7$
       $\hat{y}_R = 68.8 - .286x$

3.65 $x_L = 7.1, \ y_L = 41$
       $x_M = 11.5, \ y_M = 84.5$
       $x_R = 14.8, \ y_R = 65$
       $\hat{y}_R = 28.8 + 3.12x$ The regression equation is $\hat{y} = 27.8 + 4.66x$

There are outliers in the data that will affect the regression line. Because of the way they are situated, however, their effects will tend to counteract each other (one is on one side of the regression line and another is on the other side of the regression line). Although the y-intercepts of the two lines are close (28.8 and 27.8), their slopes are somewhat different. The regression line has a steeper slope (because of the outlier at the top of the scatterplot.) When income is 15 we have $\hat{y}_R = 75.6$ and $\hat{y} = 97.7$.

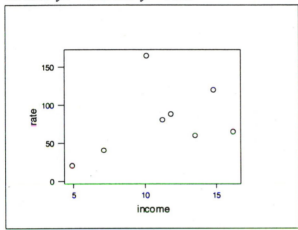

3.67   If the maintenance cost for the first register of age 3 is changed from 40 to 200 there will be a drastic change in the least squares regression line but only a slight change in the resistant line. The observation (3,200) will most definitely become an outlier and will pull the least squares line towards it. The only change to the resistant line is that y_L changes from $(35 + 40)/2 = 37.5$ to $(35 + 51)/2 = 48$. This will have minimal effect on the equation of the resistant line.

3.69   Only in case c should we use a resistant line to describe the relationship. In cases a and b there will be very little difference between the resistant line and the least squares regression line. Neither the resistant nor the regression line will explain the relationship in part d.

Section 3.6 Summary and Review

3.71   b correlated

3.73

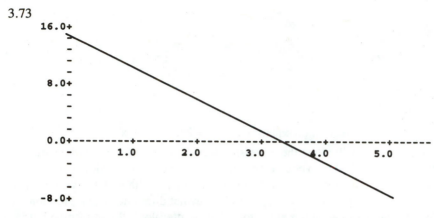

The residual for (2, 3.8) is $e = 3.8 - \hat{y}$ where $\hat{y} = 15.3 - 4.7\,(2) = 5.9$. Thus $e = 3.8 - 5.9 = -2.1$

3.75
a   140/200 = 70%
b   90/200 = 45%
c   44/58 = 75.9%
d   78/110 = 70.9%

3.77   a

	Science	Business	Education	Liberal Arts	Total
Women	9	15	22	12	58
Men	17	15	5	7	44
Total	26	30	27	19	102

b   26/102 = 25.5%
c   22/58 = 37.9%
d   22/27 = 81.5%

3.79   a

```
      16.0+        x
         -              x
Y        -        x
         -
         -
      12.0+
         -                        x
         -            x       x
         -
         -                       x
       8.0+                          x
         -
         -
         -
         -
       4.0+                                   x
         -
         --+---------+---------+---------+---------+---------+----X
           1.5       3.0       4.5       6.0       7.5       9.0
```

202

b   Correlation of X and Y = -0.959

c   The regression equation is Y = 18.2 - 1.55 X

d   The resistant line is Y = 17.53 - 1.4 X

e   There is very little difference between the regression line and the resistant line because there are no outliers and the relationship is very close to a straight line.

3.81   a

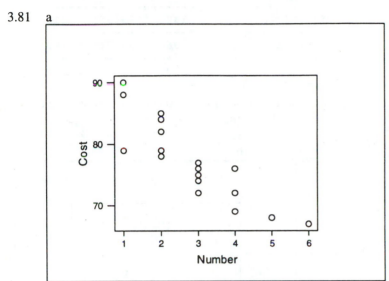

b   Correlation of Number and Cost = -0.868

c   There is a negative linear trend.

d   The regression equation is Cost = 88.6 - 4.09 Number

e   Cost per person for a family of 4 = 88.6 - 4.09 (4) = 72.24

3.83   a   The regression equation is number = - 0.03 + 1.64 age

b

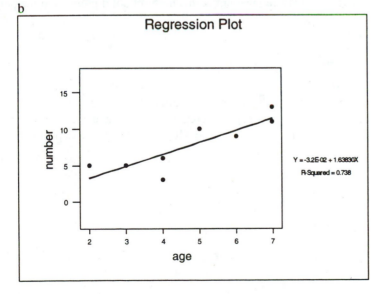

c   Correlation of age and number = 0.859

d   Strength, coordination, training

3.85

	no fatalities	at least one fatality	Total
Involved alcohol	68 (110.04)	142 (99.96)	210
no alcohol	194 (151.96)	96 (138.04)	290
Total	262	238	500

Because of the large differences between the actual counts and the expected counts we would have to say that there is an association between alcohol and fatalities.

3.87   a   b

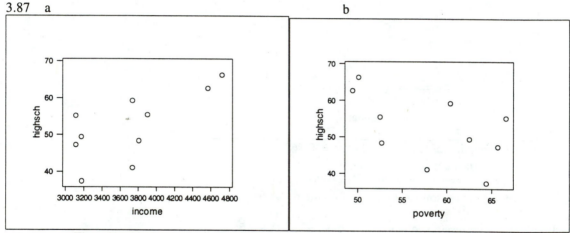

c   There is a moderate linear trend between percent with a high school diploma and the per capita income.

d   **Correlations (Pearson)**

```
          highsch    income
income    0.696
poverty  -0.542     -0.928
```

e   The correlation between percent with a high school diploma and the per capita income is 0.696. It is positive indicating that as per capita increases so does the percent with a high school diploma. There may be some discussion as to which variable is the dependent variable.

3.89   a

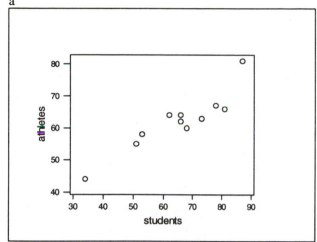

b   Correlation of students and athletes = 0.920
c   The high correlation means that there is a strong association between the graduation rates
     for all students and student athletes at schools in the Big Ten Conference.
d   No, this is just a correlation problem. We cannot say that the values of one variable depend
     on the values of the other variable.

3.91   a

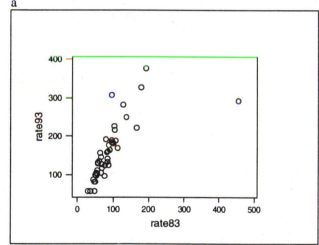

b   There is one outlier associated with Washington D. C. Because of the reclassification (see
     Note attached to table) the outlier should be removed for analysis purposes.
c   Without the outlier there is a definite linear pattern.
d   The outlier most definitely would have an effect on the regression line. It would have very
     little effect on the resistant line.
e   The regression equation is   rate93 = 87.8 + 0.778 rate83
f   The regression equation is   rate93 = 5.7 + 1.81 rate83
g   R-sq = 48.4% with all available data. R-sq = 81.1% with the one outlier removed.
     Removal of the outlier vastly improved the linear relationship.

3.93

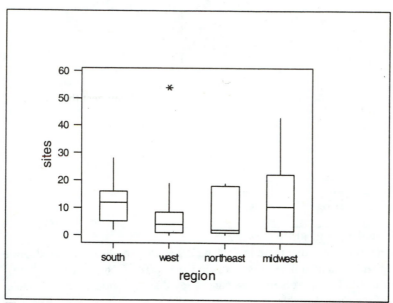

The midwest is the most variable. Except for the one outlier, the west has the least variability. The only outlier is in the west region associated with California. The median number of sites in the south and the midwest are about the same and the median number in the west and northeast are about the same. The median number of sites in the northeast is the smallest among all regions.

3.95   a   The mean annual level of Lake Victoria Nyanza is the response variable. The number of sunspots is the predictor variable.

   b

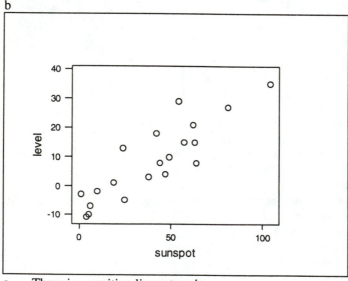

   c   There is a positive linear trend.

d   There are no outliers.
e   Pearson Correlation of level and sunspot = 0.879
f   The regression equation is   level = - 8.04 + 0.413 sunspot
g   The predicted level is 12.60 when the number of sunspots is 50. This equation should not be used to predict the level when the number of sunspots is 200 because this is way beyond the observed data. We should not extrapolate so far beyond the available data.

3.97   The median trace for the 1970 data shows a downward trend which suggest that the draft dates were not randomly distributed across the months. The median trace for the 1971 data does not show a trend. These draft dates are randomly distributed across the months.

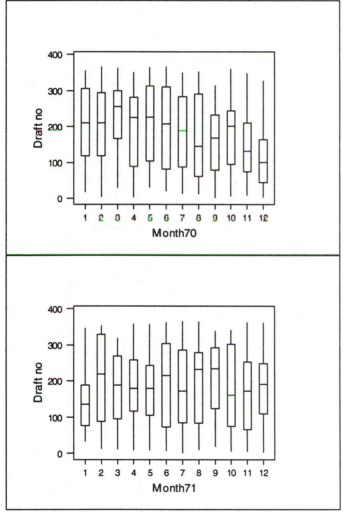

Unit 1 [Chapters 1-3]

Significant Ideas and Review Exercises

U1.1 a   T
        b   T
        c   F
        d   F
        e   F

U1.3 d   U1.5 b

U1.7
Major	Frequency
Psychology	10
Sociology	3
Criminal Just	7
Planning	2
Comp Sci	3
Other	5

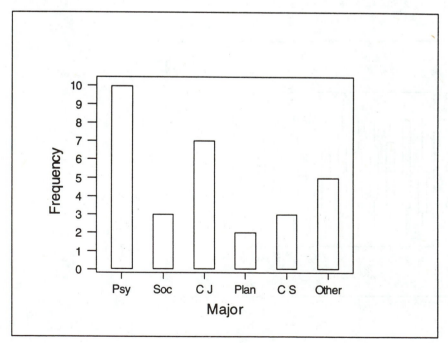

U1.9 **Stem-and-leaf of Score   N = 47**

```
Leaf Unit = 1.0

    1     3 4
    1     3
    5     3 8899
   13     4 00001111
   19     4 222223
   (9)    4 444444555
   19     4 66666677
   11     4 8
   10     5 0000
    6     5 22
    4     5
    4     5 7
    3     5
    3     6 00
    1     6 3
```

The distribution is slightly skewed right.

U1.11 a

SPORT	Relative Freq
Baseball	0.295
Basketball	0.135
Football	0.315
Golf	0.026
Ice Hockey	0.062
Soccer	0.084
Tennis	0.041
Other	0.042

b

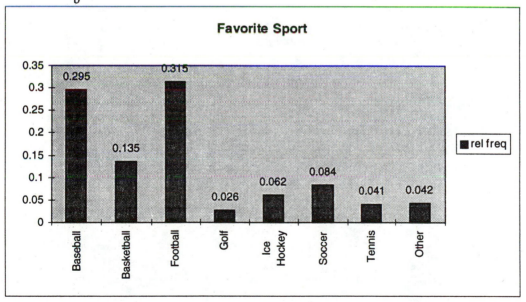

U1.13 parameter

U1.15 $\bar{y} = 57.00$ $s = 17.57$

U1.17

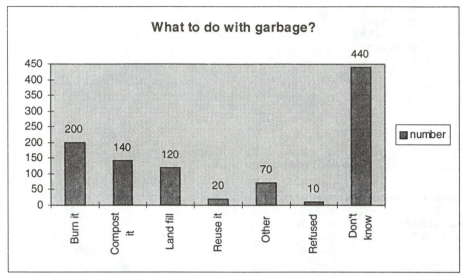

U1.19 a   Numerical b Categorical c Numerical   d Numerical
      e   Numerical f Categorical

U1.21 **Fears of 400 executives at the 1000 largest U. S. companies**

	fired	acq/merger	burnout	other	Total
1989	12	108	52	28	200
1991	44	90	24	42	200
Total	56	198	76	70	400

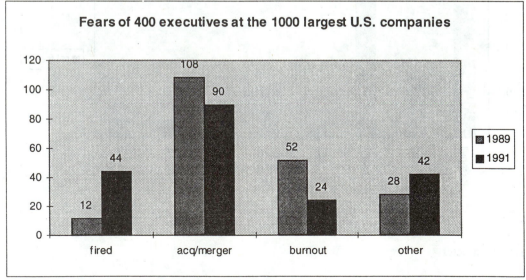

U1.23 a

```
Stem-and-leaf of rate        N  = 51
Leaf Unit = 10

    1       1 0
    3       1 33
   11       1 44445555
   21       1 6666667777
  (12)      1 888888889999
   18       2 0001111
   11       2 222233
    5       2 55
    3       2 67
    1       2 8
```

b

Letter Value Display

	DEPTH	LOWER	UPPER	MID	SPREAD
N=	51				
M	26.0	183.980		183.980	
H	13.5	164.170	214.500	189.335	50.330
E	7.0	149.830	237.290	193.560	87.460
D	4.0	141.050	255.000	198.025	113.950
C	2.5	136.085	270.730	203.408	134.645
B	1.5	121.435	278.090	199.762	156.655
	1	109.660	280.410	195.035	170.750

c

Character Boxplot

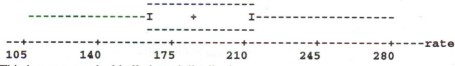

```
                             ---------------
            ---------------I     +      I---------------------
                             ---------------
       --+---------+---------+---------+---------+---------+----rate
        105       140       175       210       245       280
```

d   This is a symmetrical bell-shaped distribution.

U1.25 a

Descriptive Statistics

Variable	N	Mean	Median	TrMean	StDev	SEMean
anxiety	31	14.26	15.00	14.26	5.88	1.06

Variable	Min	Max	Q1	Q3
anxiety	3.00	27.00	9.00	19.00

b   The median, 5% trimmed mean, minimum, maximum, 1st and 3rd quartiles. SEMean is standard error of the mean and will be studied later in Chapter 7.

c
```
Stem-and-leaf of anxiety    N  = 31
Leaf Unit = 1.0

      1      0 3
      2      0 5
      4      0 66
      8      0 8899
     10      1 01
     13      1 223
    (5)      1 44555
     13      1 6677
      9      1 899
      6      2 11
      4      2 222
      1      2
      1      2 7
```

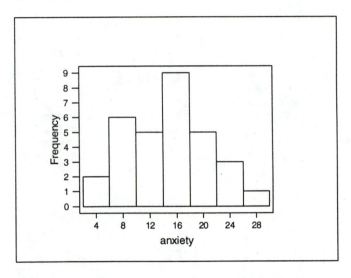

d   The data are symmetrical without unusually long tails.

U1.27 **Spearman's correlation of GPA and Exam = 0.893**
Spearman's is very close to Pearson's correlation.

U1.29 **Spearman's Correlation of X and Y = -1.000**
The outlier has no unusual effect on Spearman's correlation.

4

Probability and Probability Distributions

Section 4.1 Basic Probability Concepts

4.1 $2/6 = 1/3$

4.3 $3/10$

4.5 $25/51$

4.7 $1/3$

4.9 $1/4$

4.11 $18/38, 2/38$

4.13 $S = \{1, 2, 3, 4, 5\}$
        $A = \{3\}$ $B = \{1, 2, 3\}$ $C = \{4, 5\}$ $D = \{3,4\}$

4.15 $S = \{(H_1,M), (H_1,F), (H_2,M), (H_2,F)\}$

4.17   No

4.19   It must be .3 so that the probabilities of all outcomes sum to 1.

4.21 $S = \{(AAA),(AAD),(ADA),(DAA),(ADD),(DAD),(DDA),(DDD)\}$
        If the probability one application is approved is .6, then the outcomes are not equally. If the probability one application is approved is .5, then the outcomes are equally, and $P(\{(AAA)\})= \ldots = P(\{(DDD)\}) = 1/8$.

4.23 $S = \{(MMM),(MMF),(MFM),(FMM),(MFF),(FMF),(FFM),(FFF)\}$
        If 60% of the student body is female then the outcomes are not equally likely. If 50% of the student body is female then the outcomes are equally likely, and $P(\{(MMM)\})= \ldots = P(\{(FFF)\}) = 1/8$.

4.25   Following are the results of generating 20 random integers between 1 and 38. Your simulation may differ from these counts.

```
12 16 20   6 29 32   8   3 22 36 10 14 34 38 35 16 17   8 10   2
 r  r  b   b  b  r   b   r  b  r  b  r  b  r  r  g  b  r  b  b  b
```

Black 11/20   Red 8/20 Green 1/20

Following are the results of generating 380 random integers between 1 and 38. Again, your results will be different. In all cases, however, we should expect each number to have a count of 10. These results are typical with some numbers occurring more than 10 times and others less. We see that 21 and 36 only occurred three times each.

C1	Count						
1	7	11	7	21	3	31	12
2	10	12	10	22	15	32	11
3	10	13	8	23	8	33	11
4	10	14	13	24	10	34	6
5	10	15	14	25	10	35	13
6	17	16	12	26	8	36	3
7	11	17	14	27	14	37	8
8	8	18	12	28	10	38	4
9	11	19	12	29	12	N=	380
10	10	20	9	30	7		

4.27   a   1/20   b   12/20

Section 4.2 Event Relations and Probability Laws

4.29   a   {3,4,5,6,7}   b   {5}   c   A = {3,5,7}   d   {2,8}
       e   {4,6}   f   {3,4,5,6,7}   g   {5}   h   No, see b.

4.31   a   1-.4 = .6   b   .5+.4-.8 = .1
       c   No, A and B have an intersection, since P(A and B) = .1

4.33   Each outcome has probability ¼. P(rr) = ¼.

4.35   a   Yes. A set of golf clubs cannot be made in Denver and Phoenix and Memphis.
       b   P(M or D) = .6 + .2 = .8
       c   P(M^c) = 1-.6 = .4

4.37   a   1 - .3 = .7   b   .3 + .5 - .2 = .6   c   .2/.5 = .4   d   No, P(A|B) ≠ P(A)

4.39   Assuming independent selections for her suits, P({nn}) = (.7)(.7) =.49, P({nb}) = (.7)(.3) = .21, P({bn}) = (.3)(.7) = .21, and P({bb}) = (.3)(.3) =.09. P(two days in a row) = P(bb) = .09.

4.41   P(at least 2 are approved) = .216 + 3(.144) = 0.648

4.43   P(you stop no more than one time) = 3(.147) + .343 = .784

4.45   S has 36 equally likely outcomes. P(A) = 18/36, P(B) = 10/36, P(A and B) = 4/36, P(B|A) = 4/18

A and B are not independent because $P(B|A) \neq P(B)$. A and B are not mutually exclusive because A and B have 4 outcomes in common.

4.47 If Team A finding the hiker is independent of Team B finding the hiker, then $P(A$ and $B) = (.3)(.4) = .12$, and $P(A$ or $B) = .3 + .4 - .12 = .58$

4.49   a $39.7/.153 = 259.48$ million
       b $(259.48)*(.702) = 182.15$ million
       c $1 - .855 = .145$

4.51   a   Same as exercise 4.50 with H and N replaced by C (chipping) and N (no chipping).
       b   .016, .047, 047, .047, .141, .141, .141, .422
       c $A = \{CNN, NCN, NNC, NNN\}$, $P(A) = 3(.141) + .422 = .845$

4.53 $P(b$ and b and $b) = P(b)P(b)P(b) = (18/38)^3 = .106$

Section 4.3   Random Variables

4.55   Range of $x = \{0,1,2,3,4,5\}$, Discrete

4.57   Range of $w = \{t \mid t$ is a real number greater than $0\}$, Continuous

4.59   Range of $B = \{0,1,2,3, ---, 15\}$, Discrete

4.61   a   continuous   b   discrete   c   discrete   d   continuous   e   continuous

4.63 $P(2) = (2-1)/6 = 1/6$, $P(3) = (3-1)/6 = 2/6$, $P(4) = (4-1)/6 = 3/6$. Yes it is a probability function because the total probability $= 6/6 = 1.0$. $P(1) = 0$

4.65   a   4   b   P(even)= 222+.334=.556; P(odd)=.111+.333=.444

     c

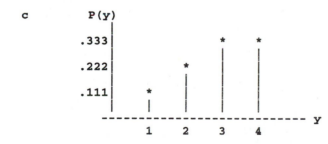

215

4.67

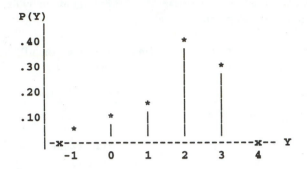

$\mu \pm 2\sigma = -.44$ to 4.04 which according to Chebyshev's Rule covers at least 75% of the distribution.

4.69

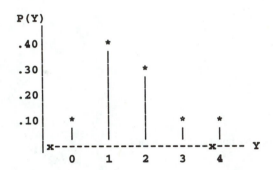

$\mu \pm 2\sigma = -.5$ to 3.9 which according to Empirical rule contains approximately 95% of the distribution.

4.71

y	P(y)	y*P(y)
-1	.05	-.05
0	.10	0.00
1	.15	.15
2	.40	.80
3	.30	.90

$\mu = -.05 + 0 + .15 + .80 + .90 = 1.8$

4.73 $\mu = 0 + .4 + .6 + .3 + .4 = 1.7$

4.75

y	P(y)	y*P(y)	y^2*P(y)
-1	.05	-.05	.05
0	.10	0.00	0.00
1	.15	.15	.15
2	.40	.80	1.60
3	.30	.90	2.70

$\sigma^2 = .05 + 0 + .15 + 1.6 + 2.7 - (1.8)^2 = 1.26$ $\sigma = \sqrt{1.26} = 1.12$

4.77 $\sigma^2 = 0 + .4 + 1.2 + .9 + 1.6 - (1.7)^2 = 1.21 \quad \sigma = \sqrt{1.21} = 1.1$

4.79 Follow are the results obtained with the Tally command in Minitab. Your results will be different.

Summary Statistics for Discrete Variables

C1	Count		C2	Count		C3	Count		C4	Count
0	5		0	5		0	5		0	5
N=	5		N=	5		N=	5		N=	5

C5	Count		C6	Count		C7	Count		C8	Count
0	4		0	3		0	5		0	5
1	1		1	2		N=	5		N=	5
N=	5		N=	5						

C9	Count		C10	Count		C11	Count		C12	Count
0	5		0	5		0	5		0	5
N=	5		N=	5		N=	5		N=	5

C13	Count		C14	Count		C15	Count		C16	Count
0	3		0	4		0	3		0	5
1	2		1	1		1	2		N=	5
N=	5		N=	5		N=	5			

C17	Count		C18	Count		C19	Count		C20	Count
0	5		0	5		0	5		0	4
N=	5		N=	5		N=	5		1	1
									N=	5

C21	Count		C22	Count		C23	Count		C24	Count
0	4		0	5		0	5		0	5
1	1		N=	5		N=	5		N=	5
N=	5									

C25	Count		C26	Count		C27	Count		C28	Count
0	4		0	3		0	4		0	4
1	1		1	2		1	1		1	1
N=	5		N=	5		N=	5		N=	5

C29	Count		C30	Count		C31	Count		C32	Count
0	5		0	5		0	5		0	4
N=	5		N=	5		N=	5		1	1
									N=	5

Following is the summary of x = number of females selected out of 5 for 32 simulations.

x	rel freq	
0	20/32 =	.625
1	8/32 =	.250
2	4/32 =	.125
3	0/32 =	0
4	0/32 =	0
5	0/32 =	0

4.83

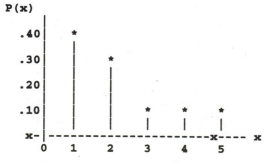

217

$\mu \pm 2\sigma$ = -.46 to 4.86 which according to Chebyshev's Rule covers at least 75% of the distribution.

Section 4.4 The Binomial Distribution

4.85   Using the Binomial probability tables
      a   .201   b .147   c .312 d .069

4.87   a $\mu = (10)(.8) = 8$, $\sigma = \sqrt{(10)(.8).2)} = 1.2649$
        b $\mu = (15)(.3) = 4.5$, $\sigma = \sqrt{(15)(.3)(.7)} = 1.7748$
        c $\mu = (6)(.5) = 3$, $\sigma = \sqrt{(6)(.5)(.5)} = 1.2247$
        d $\mu = (4)(.9) = 3.6$, $\sigma = \sqrt{(4)(.9)(.1)} = .6$

4.89   Work it out using the binomial probability formula or use a computer.

4.91   Let Y be the number of approved applications out of 10. Then Y is a binomial random variable with n = 10 and $\pi = .6$
    a   Using the Binomial probability tables, .121 + .040 + .006 = .167
    b $\mu = (10)(.6) = 6$
    c $\sigma = \sqrt{(10)(.6)(.4)} = 1.549$

4.93   Let Y be the number favoring the action in a sample of 15 congressman. Then Y is a binomial random variable with n = 15 and $\pi = .5$.
    a   Using the Binomial probability tables, .153+.092+.042+.014+.003 = .304
    b $\mu = (15)(.5) = 7.5$

4.95   a   Having an accident during the week could be identified as a "success"; however, the number of trials is not specified. So the number of accidents per week is not a binomial random variable.
        b   A violent crime being committed during a month could be identified as a success; however, the number of trials is not specified. So the number of crimes committed per month is not a binomial random variable.
        c   If we assume that the probability of a successful heart transplant is the same for all 5 patients, then the number of successful heart transplant is a binomial random variable with n = 5.
        d   The length of a prison term is continuous and cannot be a binomial random variable.
        e   The number approved for food stamps is binomial with n = 50.

4.97   Let Y be the number of blacks selected on a 12 person jury. Then Y is a binomial random variable with n = 12 and $\pi = .6$. P(4 or fewer blacks selected from 12)=.042 + .012 + .002= .056. Because this probability is small, it is doubtful that the jury was selected randomly.

4.99

y	P(y)
0	$(.75)^3 = .421875$
1	$3(.25)(.75)^2 = .421875$
2	$3(.25)^2(.75) = .140625$
3	$(.25)^3 = .015625$

$\mu = (3)(.25) = .75$, $\sigma = \sqrt{(3)(.25)(.75)} = .75$

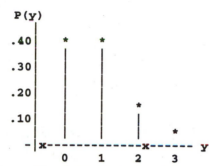

$\mu \pm 2\sigma = -.75$ to 2.25 which according to Chebyshev's Rule covers at least 75% of the distribution.

4.101 Following are the results of 30 simulations where n = 15 trials and $\pi = .70$. Your results will be different.

11, 11, 11, 8, 10, 7, 11, 12, 12, 8, 8, 11, 9, 9, 9, 10, 10, 9, 10, 14, 8, 13, 10, 8, 10, 7, 12, 11, 9, 11

In no simulation did all 15 have a private insurance plan. Out of 15 we would expect $\mu = (15)(.7) = 10.5$ to have a private insurance plan. The number 11 came up seven times.

4.103 Following are the results of 20 simulations where n = 50 trials and $\pi = .56$. Your results will be different

24, 33, 28, 34, 35, 30, 26, 32, 27, 25, 29, 22, 26, 24, 27, 33, 34, 28, 28, 37

Out of a sample of 50 we would expect $\mu = (50)(.56) = 28$ to be white. In the 20 simulations 28 came up three times.

4.105 Let Y be the number who experience relief out of 16. Y is binomial with n = 16 and $\pi = .9$. Using the binomial probability tables,
   a $P(Y = 16) = .185$
   b $P(Y \geq 14) = .274 + .329 + .185 = .788$
   c $\mu = 16(.9) = 14.4$

Section 4.5 The Normal Distribution

4.107 a   .9773 or 97.73%   b   .9953 or 99.53%
       c   .9131 or 91.31%   d   .0228 or 2.28%
       e   .8166 or 81.66%   f   .2786 or 27.86%

4.109 $Z = (X - 50)/10$
  a $P(40 \leq X \leq 56) = P(-1 \leq Z \leq 0.6) = .7257 - .1587 = .5670$
  b $P(X \geq 64) = P(Z \geq 1.4) = .1 - .9192 = .0808$

4.111 $Z = (T - 30)/10$ $P(T > 60) = P(Z > 3) = 1 - .9987 = .0013$

4.113 Its mean and standard deviation (or variance).

4.115 a   Because weight can be any value greater than 0, W is continuous.
  b   No, because the distribution for W is not approximately normal.
  c   No, because the distribution for W is not approximately normal.
  d   200 and 300
  e   The proportion of players that weigh between 200 and 250.
  f   Yes, the weights of football players typically fall into two groups. The lineman are heavier than others players.

4.117 Let Y be the score on the test. Y is normal with $\mu = 450$ and $\sigma = 100$, $Z = (Y - 450)/100$, and Y $= 450 + Z(100)$.

  a $P(350 < Y < 550) = P(-1 < Z < 1) = .8413 - .1587 = .6826$
  b $P(Y > 400) = P(Z > -.5) = 1 - .3085 = .6915$
  c   The upper 5% would be the 95th percentile. So $z = 1.64$ and $y = 450 + 1.64(100) = 614$.

4.119 Let Y be the length of time required to run the maze. Y is normal with $\mu = 15$, $\sigma = 3$, $Z = (Y - 15)/3$, and $Y = 15 + Z(3)$.

  a $P(10 < Y < 20) = P(-1.67 < Z < 1.67) = .9525 - .0475 = .9050$
  b   For the 10th percentile $z = -1.28$, so $y = 15 - 1.28(3) = 11.16$
  c   For the 90th percentile $z = 1.28$, so $y = 15 + 1.28(3) = 18.84$

4.121 Let Y the contamination level index. Y is normal with $\mu = 160$ and $\sigma = 20$. $Z = (Y - 160)/20$

  a $P(Y > 190) = P(Z > 1.5) = 1 - .9332 = .0668$
  b $P(150 < Y < 170) = P(-.5 < Z < .5) = .6915 - .3085 = .3830$
  c $Y = 160 + Z(20)$. The 80th percentile for Z is $z = .84$. So $Y = 160 + (.84)(20) = 176.8$ is the level at which the beach should be closed.

4.123 The one extreme observation on the right tail causes the pattern of the probability plot to indicate that the distribution is skewed right. Without that observation the data could possibly be normally distributed.

4.125

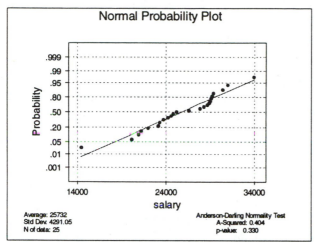

The normal probability plot suggests that the data are coming from a normally distributed population. This does not agree with the assessment in Exercise 2.100. This information along with the analysis in Exercise 2.100 should be used to classify this distribution.

4.127

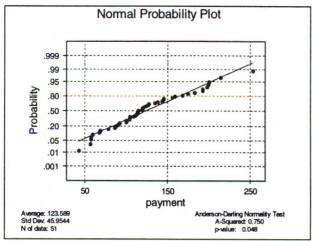

Except for the two observations on the right of the graph the pattern is that of a distribution that is short tailed. The analysis in Exercise 2.110 suggest that the distribution is slightly skewed right because of the outlier on the right tail. Without the outlier the distribution could be classified as symmetrical without excessively long tails.

4.129   The lower quartile q_1 is the 25th percentile and therefore is $-.675\sigma$ from the mean. The upper quartile q_3 is the 75th percentile and therefore is $+.675\sigma$ above the mean. The q-spread is then the difference, q-spread $= +.675\sigma - (-.675\sigma) = 1.35\sigma$. The lower eighth e_1 is the 12.5^{th} percentile and therefore is -1.15σ from the mean and e_7 is $+1.15\sigma$ above the mean. Thus we have

e-spread $= 1.15\sigma - (-1.15\sigma) = 2.3\sigma$.

Section 4.6 Summary and Review

4.131 d   4.133 d   4.135 a

4.137 a   S = {rrr, rrn, rnr, nrr, rnn, nrn, nnr, nnn}
       b   No, only 25% return for an advanced degree. If 50% return then the outcomes are equally
            likely.
       c $(.75)^3 = .422$
       d $3(.25)^2(.75) + (.25)^3 = .156$

4.139 1/3, 0

4.141 a   3/5   b   3/7

4.143 a   BC, BM, BP, BS, CM, CP, CS, MP, MS, PS
       b   1/10   c   4/10   d   3/10

4.145 .4452, .2119

4.147 a   T   b F   c F   d T   e F   f F

4.149 a   .0359   b   3.14

4.151 a $z = 3.33, .0004$
       b $z = .83, .7967$
       c $z = 1.67$ and $z = 5$, $1 - .9525 = .0475$

4.153 a $z = -1.33$ and $z = .6$, $.7257 - .0918 = .6339$
       b $z = 1.6$, $1 - .9452 = .0548$
       c $50 + 1.28(15) = 69.2$

4.155 a   L = {(1,1),(6,6)}; P(L) = 2/36
       b $1 - 2/36 - 8/36 = 26/36$

4.157 S = {RR, RG, RB, GR, GG, GB, BR, BG, BB}
       A = {RR}
       B = {GG, GB, BG, BB}
       C = {RG, RB, GR, BR}

4.159 a   S = {(1,H),(1,T),(2,1),(2,2),(2,3),(2,4),(2,5),(2,6)
            (3,H),(3,T),(4,1),(4,2),(4,3),(4,4),(5,5),(4,6)
            (5,H),(5,T),(6,1),(6,2),(6,3),(6,4),(6,5),(6,6)}
       Let O be the event an odd number occurs, E be the event an even number occurs, H be the event
       a head occurs, and T be the event a tail occurs.
       $P\{(O,H)\} = P(O)P(H) = (1/6)(1/2) = 1/12$ for O=1,3,5

$P\{(O,T)\} = P(O)P(T) = (1/6)(1/2) = 1/12$ for O=1,3,5

$P\{(E,x))\} = P(E)P(\{x\}) = (1/6)(1/6) = 1/36$, for E=2,4 6 and x=1,2,3,4,5,6

b   P(head on coin) = 3(1/12) = 1/4

4.161   Let Pi = Pair i, for i = 1,2,3,4.

S = {{P1,P2},{P1,P3},{P1,P4},{P2,P3},{P2,P4},{P3,P4}}

Because she randomly chooses 2 pairs, the outcomes will be equally likely and each will have probability 1/6.

4.163   Let Y be the number of completed passes out of the next 10. Y is binomial with n = 10 and π = .6

a   P(Y = 7,8, or 9) = .215 + .121 + .040 = .376

b $\mu = 10(.6) = 6$

c $\sigma = \sqrt{10(.6)(.4)} = 1.55$

4.165   Let Y be the number of children out of the next 4 calls. Y is binomial with n = 4 and π = .64.

$P(Y \geq 2) = 6(.64)^2(.36)^2 + 4(.64)^3(.36) + (.64)^4 = .864$

4.167   Gain = [($9)(.01) + (-$1)(.99)]100 = -$90

4.169   Your simulation will yield similar but different results.

Summary Statistics for Discrete Variables

C3	Count	Percent
2	7	3.50
3	7	3.50
4	13	6.50
5	26	13.00
6	33	16.50
7	34	17.00
8	33	16.50
9	17	8.50
10	13	6.50
11	8	4.00
12	9	4.50
N=	200	

From the simulation we recommend that the probability of rolling a total of 7 to be 34/200 = 17% which is very close to the theoretical probability of 6/36 = 16.67%.

4.171   Your simulation will yield similar but different results.

Summary Statistics for Discrete Variables

C1	Count
1	34
2	43
3	43
4	40
5	40
N=	200

The integers 1-5 are evenly distributed as they should be. Red came up 117/200 = 58.5% which is very close to the theoretical 60%.

4.173 Your simulation will yield similar but different results.

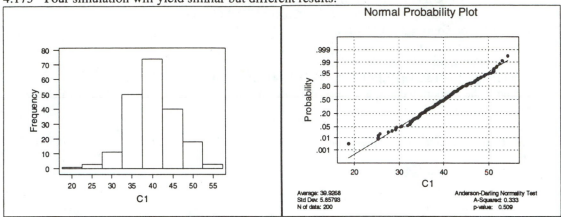

The histogram has a bell shaped appearance. With the exception of one outlier. the normal probability plot indicates that the data are normally distributed.

4.175 a, b, and c

2	3	switch
3	1	switch
1	2	switch
2	2	
3	2	switch
3	2	switch
3	2	switch
2	2	
2	3	switch
1	1	
2	1	switch
2	1	switch
2	2	
3	2	switch
3	1	switch
1	3	switch
2	2	
3	3	
3	3	
1	2	switch
2	3	switch
1	2	switch
1	2	switch
3	1	switch
3	1	switch
3	1	switch
3	1	switch
2	1	switch
3	1	switch
1	1	

    d   switch 22/30 stay 8/30

    e   Based on this simulation, the contestant should switch.

5

Sampling Distributions

Section 5.1 Sampling Distributions

5.1 M - the sample median

5.3 M_1-M_2 - the difference between two sample medians

5.5 $\bar{y}$ - the sample mean

5.7 The figure $1.22 is a statistic because it was found from a sample of 6000 service stations.

5.9

Sample	Prob	Sample	Prob	Sample	Prob
(1,1,1)	1/27	(1,2,1)	1/27	(1,6,1)	1/27
(1,1,2)	1/27	(1,2,2)	1/27	(1,6,2)	1/27
(1,1,6)	1/27	(1,2,6)	1/27	(1,6,6)	1/27
(2,1,1)	1/27	(2,2,1)	1/27	(2,6,1)	1/27
(2,1,2)	1/27	(2,2,2)	1/27	(2,6,2)	1/27
(2,1,6)	1/27	(2,2,6)	1/27	(2,6,6)	1/27
(6,1,1)	1/27	(6,2,1)	1/27	(6,6,1)	1/27
(6,1,2)	1/27	(6,2,2)	1/27	(6,6,2)	1/27
(6,1,6)	1/27	(6,2,6)	1/27	(6,6,6)	1/27

a

$\bar{y}$	1	4/3	5/3	2	8/3	3	10/3	13/3	14/3	6
$P(\bar{y})$	1/27	3/27	3/27	1/27	3/27	6/27	3/27	3/27	3/27	1/27

b

M	1	2	6
P(M)	7/27	13/27	7/27

225

c

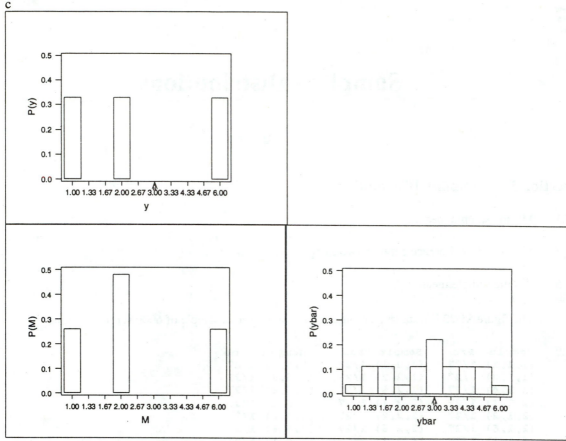

d   The sampling distribution of $\bar{y}$ is centered at μ but the sampling distribution of the median M is center at 2.778. The sampling distribution of the median M is more spread out than is the sampling distribution of $\bar{y}$.

e   Because the sampling distribution of $\bar{y}$ is centered at μ and is less variable, we should choose it as the better measure of the center of the distribution.

5.11 a   347
      b   p - the sample proportion
      c   .03/2 = .015
      d   very possible
      e   31% to 37%

5.13 The population of interest is all Americans. Because the figure 50% pertains to all Americans it is a parameter.

Section 5.2 The Sampling Distributions of $\bar{y}$

5.15   The standard deviation of $\bar{y} = \sigma / \sqrt{n}$. We generally expect that $\bar{y}$ will be within $\pm 2\sigma / \sqrt{n}$ of μ.

5.17   a $2\sigma / \sqrt{n} = 2(100 / \sqrt{200}) = 14.14$, No, because 710 is greater than $690.8 + 14.14 = 704.94$.
      b $2\sigma / \sqrt{n} = 2(200 / \sqrt{200}) = 28.28$, Yes, because 710 is less than $690.8 + 28.28 = 719.08$.
      c   Rarely would $\bar{y}$ exceed $\mu + 3\sigma / \sqrt{n} = 690.8 + 3(100/\sqrt{200}) = 712.01$

5.19   Let the population represent the length of a field goal. $\mu = 38.2$ and $\sigma = 6.4$. With $n = 41$, $\bar{y}$ is approximately normal with mean 38.2 and standard deviation $6.4/\sqrt{41} = 1$. So $Z = (\bar{y} - 38.2)/1$
     $P(\bar{y} < 37) = P(Z < -1.2) = .1151$

5.21   a $Z = (30,000 - 28,500)/(2,400 = .625$ $P(y > 30,000) = P(Z > .625) = 1 - .7357 = .2643$
      b $Z = (30,000 - 28,500)/(2,400/\sqrt{36}) = 3.75$ $P(\bar{y} > 30,000) = P(Z > 3.75) = 1 - .9999 =$
        $.0001$

5.23   Let Y be the starting salary.
      a   Y is normal and $Z = (Y - 48,000)/4000$
                $P(Y > 53,000) = P(Z > 1.25) = .1056$
      b $\bar{y}$ is normal and $Z = (\bar{y} - 28,000)/1000$
       $P(\bar{y} > 53,000) = P(Z > 5) = .0000003$
       So it would be very unusual.

5.25 $Z = (1900 - 1800)/(400/\sqrt{90}) = 2.37$
     $P(\bar{y} > 1900) = P(Z > 2.37) = 1 - .9911 = .0089$

5.27   a $Z = (34 - 31)/2 = 1.5$, $Z = (28 - 31)/2 = -1.5$
       probability $= .9332 - .0668 = .8664$
      b $Z = (34 - 31)/(2/\sqrt{25}) = 7.5$, $Z = (34 - 31)/(2/\sqrt{25}) = 7.5$
       probability $= 1^- - 0 = 1^-$, almost certain.

5.29   a $50/\sqrt{64} = 6.25$ b $50/\sqrt{400} = 2.5$

5.31 $Z = (99.5 - 99)/(2/\sqrt{50}) = 1.77$, $Z = (98.5 - 99)/(2/\sqrt{50}) = -1.77$
     Probability of a false alarm $= 2(1 - .9616) = .0768$.

5.33

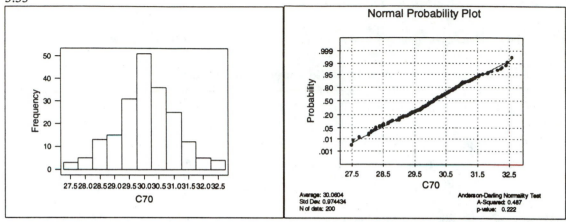

The histogram in Exercise 5.32 is close to being normally distributed, centered at 30, and ranges from about 25 to 35. The histogram in this exercise is also close to normal, centered at 30, but ranges from about 27.5 to 32.5. The normal probability plot here also support the conjecture that the sampling distribution of $\bar{y}$ is normally distributed when the population is also normally distributed. The only difference between the two exercises is that this distribution is less variable. This is a result of the fact that the standard deviation of the sampling distribution of $\bar{y}$ is $\sigma/\sqrt{n} = 8/\sqrt{64} = 1$, whereas for the previous exercise, when n = 16, the standard deviation of the sampling distribution of $\bar{y}$ was $8/\sqrt{16} = 2$.

Descriptive Statistics

Variable	N	Mean	Median	TrMean	StDev	SEMean
C70	200	30.060	30.106	30.061	0.974	0.069

Variable	Min	Max	Q1	Q3
C70	27.484	32.580	29.458	30.689

Notice that the mean of the distribution of sample means is 30.060, very close to 30, and the standard deviation is 0.974, very close to 1.

5.35 b

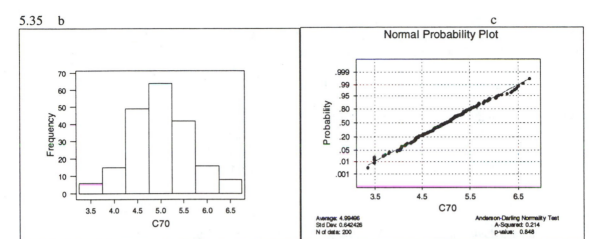

The histogram is closely approximated by a normal distribution. The normal probability plot supports the conjecture that the sampling distribution of $\overline{y}$ is approximately normally distributed.

d   Descriptive Statistics

Variable	N	Mean	Median	TrMean	StDev	SEMean
C70	200	4.9950	4.9912	4.9903	0.6424	0.0454

Variable	Min	Max	Q1	Q3
C70	3.3442	6.7374	4.5622	5.3868

The mean of the distribution of sample means is 4.995 which is very close, as it should be, to the population mean of 5. The standard deviation of the distribution of sample means, 0.6424, is smaller than it was in Exercise 5.34, 1.2869, when the sample size was only 16.

e   In comparing the results of this exercise and the results found in Exercise 5.34 we see that when the sample size is only 16, the sampling distribution of $\overline{y}$ is not approximately normally distributed. When the sample size is 64, however, the sampling distribution of $\overline{y}$ is approximately normally distributed. This is a result of the Central Limit Theorem that says that it will be approximately normally distributed when the sample size is relatively large.

Section 5.3 Sampling Distribution of the Sample Proportion p

5.37   A Bernoulli population is one in which there are only two possible outcomes that are usually designated as success and failure. The probability that an outcome is a success is denoted as π and the probability of failure is then $1 - \pi$. Examples include a population of voters where each person is for or against an issue and a population of homes that either heat with gas or they do not.

5.39   a   mean = .7, standard deviation $=\sqrt{(.7)(.3)/100} = .0458$

       b   mean = .7, standard deviation $=\sqrt{(.7)(.3)/400} = .0229$

       c   mean = .7, standard deviation $=\sqrt{(.7)(.3)/1000} = .0145$

       d   mean = .7, standard deviation $=\sqrt{(.7)(.3)/1600} = .0115$

5.41   The maximum standard deviation of p occurs when $\pi = .5$.

       a $\sqrt{(.5).5)/25} = .1$ b $\sqrt{(.5).5)/100} = .05$

       c $\sqrt{(.5).5)/200} = .0354$ d $\sqrt{(.5).5)/500} = .0224$

       e $\sqrt{(.5).5)/1000} = .0158$ f $\sqrt{(.5).5)/2000} = .0112$

5.43 $\pi = .87$ and n = 400. By the Central Limit Theorem p is approximately normal with mean = .8 and

     standard deviation $= \sqrt{(.87)(.13)/400} = .017$. Thus, $Z = (p - .87)/.017$.

       a $P(p > .88) = P(Z > .59) = 1 - .7224 = .2776$

       b $P(p < .85) = P(Z < -1.18) = .1190$

       c $P(.80 < p < .85) = P(-4.12 < Z < -1.18) = .1190^-$

5.45   In a random sample of 1000, by the Central Limit Theorem p is approximately normal with mean π and standard deviation $\sqrt{\pi(1-\pi)/1000}$. Using p = .23 as an estimate of π, we would estimate the standard deviation of p to be $\sqrt{(.23)(.77)/1000} = .0133$ and we would expect p to be no more than 2 standard deviations from the true value of π, i.e. we would expect p to be within $\pm 2(.0133) = \pm .0266$ of the true value of π.

5.47 $\pi = .7$ and n = 200. By the Central Limit Theorem p is approximately normal with mean = .7 and standard deviation $= \sqrt{(.7)(.3)/200} = .0324$ and $Z = (p - .7)/.0324$.

     $P(p > .65) = P(Z > -1.54) = 1 - .0618 = .9382$

5.49 $\pi = .41$ and n = 50. By the Central Limit Theorem p is approximately normal with mean = .41 and standard deviation $= \sqrt{(.41)(.59)/50} = .07$ and $Z = (p - .41)/.07$.

       a   .41

       b $\sqrt{(.41)(.59)/50} = .07$

       c   Because the sample size is four times as large, the standard deviation will be half as much, i. e. $\sqrt{(.41)(.59)/200} = .035$

       d $P(p > .5) = P(Z > 1.29) = 1 - .9015 = .0985$

       e $P(p > .5) = P(Z > 2.57) = 1 - .9949 = .0051$

       f   The difference is due to the smaller standard deviation that results from a larger sample size.

5.51 $\pi = .59$ and n = 400. By the Central Limit Theorem p is approximately normal with mean = .59 and standard deviation $= \sqrt{(.59)(.41)/400} = .0246$ and $Z = (p - .59)/.0246$.

       a $P(p > .6) = P(Z > .41) = 1 - .6591 = .3409$

       b $\pi = .43$, $Z = (p - .43)/\sqrt{(.43)(.57)/400}$, $P(p > .6) = P(Z > 6.87) = 0^+$

       c   The answer to part b is so small because with only 43% voting for Clinton, it is very unlikely that more than 60% of a random sample of 400 voters would vote for him. The difference between 60% and 43% is much greater than the difference between 60% and 59%.

5.53 Assuming $\pi = .5$, with n = 1008, by the Central Limit Theorem p is approximately normal with mean = .5 and standard deviation = $\sqrt{(.5)(.5)/1008}$ = .016 and Z = (p -.5)/.016.
$P(p \le .44) = P(Z \le -3.75) = 0^+$ Because the probability that it happened by chance is so small, there is strong evidence that the aide is in error.

5.55 $\pi = .25$ and n = 36, $\sqrt{(.25)(.75)/36}$ = .0722 and Z = (p -.25)/.0722
$P(p > .3056) = P(Z > .77) = 1 - .7794 = .2206$ These results are not unusual and therefore are not substantially better than what would normally be expected.

Section 5.4 Sampling Distributions of Other Useful Statistics

5.57   a   The sampling distributions of all three statistics are centered at .5.
       b   When the population is uniformly distributed, all three sampling distributions will be centered at the mean of the population.
       c   The median has the greatest standard deviation and the midrange has the least standard deviation
       d   The midrange provides the best estimate of the population mean.

5.59 $\sqrt{\sigma_1^2/n_1 + \sigma_2^2/n_2} = \sqrt{(.4)^2/40 + (.4)^2/45} = .4\sqrt{1/40 + 1/45} = .087$ The sampling distribution of $\overline{y}_1 - \overline{y}_2$ is centered at 14 - 14 = 0 and is approximately normally distributed.

5.61 The standard deviation of $p_1 - p_2$ is $\sqrt{\pi_1(1-\pi_1)/n_1)+(\pi_2(1-\pi_2)/n_2)} = \sqrt{(.15)(.85)/250+(.16)(.84)/300}$ = .031 The sampling distribution is centered at $\pi_1 - \pi_2 = -.01$ and is approximately normally distributed.

Section 5.5 Summary and Review

5.63 If we are only interested in automobile accidents in 1993 then it is a parameter. If the population is all automobile accidents then these data are assumed to be a sample and then 15.9 is a statistic.

5.65 p = 40/200 = .2, Z = (.2 - .176)/$\sqrt{(.176)(.824)/200}$ = .89 $P(p > .2) = P(Z > .89) = 1 - .8133$ = .1867

5.67   a   10, .2   b 10, .4   c 50, 1.2   d 50, 2.4

5.69   The sampling distribution of $\bar{y}$ should be approximately normal, so the standard deviation of $\bar{y}$ can be approximated with Range/4. i.e., standard deviation is approximately $800/4 = 200$.

5.71 $Z = (60 - 56.7)/(9.3/\sqrt{36}) = 2.13$ $P(\bar{y} > 60) = P(Z > 2.13) = 1 - .9834 = .0166$

5.73   For sufficiently large n, the sampling distribution of p is approximately normally distributed with a mean of π and a standard deviation of $\sqrt{\pi(1-\pi)/n}$.

5.75 $\sqrt{(.5)(.5)/200} = .0354$

5.77   a $Z = (145 - 149)/(40/\sqrt{100}) = -1$ $P(\bar{y} > 145) = P(Z > -1) = 1 - .1587 = .8413$
       b $Z = (155 - 149)/(40/\sqrt{100}) = 1.5$ $P(\bar{y} < 155) = P(Z < 1.5) = .9332$
       c $Z = (159.6 - 149)/(40/\sqrt{100}) = 2.65$ $Z = (151.2 - 149)/(40/\sqrt{100}) = .55$
           $P(151.2 < \bar{y} < 159.6) = P(.55 < Z < 2.65) = 9960 - .7088 = .2872$

5.79   a $Z = (80 - 75)/15 = .33$ $P(y > 80) = P(Z > .33) = 1 - .6293 = .3707$
       b $Z = (95 - 75)/15 = 1.33$ $Z = (84 - 75)/15 = .6$ $P(84 < y < 95) = P(.6 < Z < 1.33) = .9082 - .7257 = .1825$
       c $75 + (.84)(15) = 87.6$
       d $Z = (80 - 75)/(15/\sqrt{25}) = 1.67$ $P(\bar{y} > 80) = P(Z > 1.67) = 1 - .9525 = .0475$

5.81   standard deviation $= 2.5/\sqrt{n} = .4$, so n $= (2.5/.4)^2 = 39.0625 \approx 40$.

5.83   Let Y be the index score. $\bar{y}$ is approximately normal and $Z = (\bar{y} - 45)/1.35$
       $P(\bar{y} < 43) = P(Z < -1.48) = .0694$

5.85 $\pi = .12$, p is approximately normal and $Z = (p - .12)/.0257$ $P(p < .094) = P(Z < -1.01) = .1562$

5.87 $Z = (p - .29)/\sqrt{.29(.71)/200}$ If $p = 65/200 = .325$ then $Z = (.325 - .29)/\sqrt{.29(.71)/200} = 1.09$ which is not unusual at all. It is very conceivable that more than 65 out of a random sample of 200 criminal victimizations would involve firearms.

5.89   standard deviation $= 15/\sqrt{n} = 2$ thus n $= (15/2)^2 = 56.25 \approx 57$

5.91 $Z = (9.2 - 8)/(3/\sqrt{100}) = 4$. The amount these students study is a full 4 standard deviations above what typical college students study. This almost certainly did not happen by chance.

5.93　a

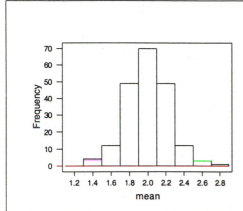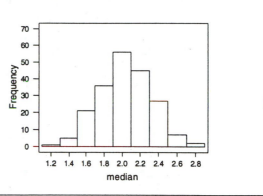

b　Both sampling distributions are centered at 2.0.

c　Both distributions are symmetrical and appear to be normally distributed.

d　The distribution of the median is more variable.

e　The values for the sample mean are more tightly clustered about the population mean than are the values of the sample median when the population is normally distributed.

5.95　a

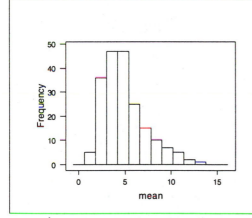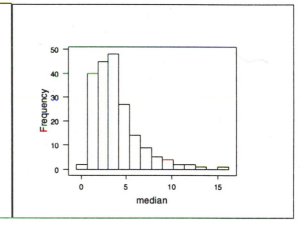

b　**Descriptive Statistics**

Variable	N	Mean	Median	TrMean	StDev	SEMean
mean	200	4.922	4.536	4.749	2.324	0.164
median	200	3.811	3.249	3.559	2.524	0.178

Variable	Min	Max	Q1	Q3
mean	1.238	13.562	3.195	5.931
median	0.508	15.372	1.967	4.752

Using the describe command we see that the sampling distribution of the mean is centered around 5 (the mean of the population) and the sampling distribution of the median is centered around 3.8.

233

c   Both distributions are skewed right.

d   From the descriptive statistics we see that the distribution of the median is a little more variable than is the distribution of the mean.

e   The sampling distributions are skewed right because the original population (exponential) is skewed right. These sampling distributions are based on averages of 5 observations from the population, consequently they are not as skewed as the population.

5.97 **Tabulated Statistics**

    ROWS: gender   COLUMNS: smoke

	0	1	ALL
0	375	105	480
1	418	102	520
ALL	793	207	1000

    CELL CONTENTS --
           COUNT

a   Using the Cross Tabulation command we find that 207/1000 = 20.7% of the kids smoke.

b   From the tabulated statistics we find that 105 of the 480 females smoke. The percent is 105/480 = 21.875%

c   We find that 102 of the 520 males smoke. The percent is 102/480 = 21.25%

d   There are 520 males and 480 females in the study.

e   Slightly more females (21.875%) smoke than males (21.25%).

f   These percents are considerably higher than those found in the NCHS survey.

6

Describing Distributions

Section 6.1 Data Analysis

6.1 Based on a midsummary analysis we should classify this distribution as skewed right because the midsummaries get progressively larger as we scan down from the median to the midrange.

6.3 a The dotplot shows two separate distributions, one centered around 27 and the other centered around 45. This was not discovered with either the midsummary analysis or the boxplot.

b One mode is for cars and the other is for trucks.

c The average mpg from the data is 36.85. This is not representative of the data because no data are close to this figure. The closest observation is 41

d These data should be analyzed as two separate data sets—one for cars and one for trucks.

6.5 a **Character Stem-and-Leaf Display**

```
Stem-and-leaf of miller    N  = 25
Leaf Unit = 1.0

     1      1 4
     2      1 6
     5      1 889
     7      2 01
    12      2 22233
    (5)     2 45555
     8      2 67
     6      2 9
     5      3 01
     3      3 3
     2      3 45
```

b

	DEPTH	LOWER	UPPER	MID	SPREAD
N=	25				
M	13.0		24.000	24.000	
H	7.0	21.000	27.000	24.000	6.000
	1	14.000	35.000	24.500	21.000

c The midsummaries are all about the same, we should classify the shape as being symmetrical.

d   **Character Boxplot**

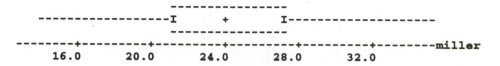

e   · The boxplot also looks symmetrical. We should classify the parent distribution as being symmetrical.

6.7   The normal probability plot indicates no significant departures from normality. In Exercise 6.5 we concluded that the distribution is symmetrical, we can now say that it is also normally distributed.

6.9   The normal probability plot pattern is that of a skewed right distribution. This agrees with the assessment given in Exercises 6.6 and 6.8.

6.11   a   b

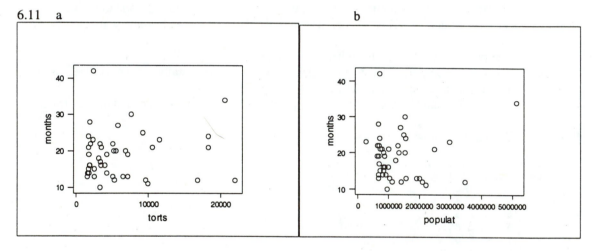

c

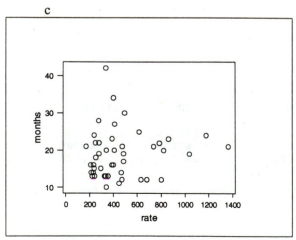

d   There is a slight relationship between the number of months to process a tort and the population size of the county. As the population size increases the processing time decreases.

6.13   The histogram shows that the waiting times are bimodal. Normally one would expect a symmetrical unimodal distribution. The mean wait time is 72.314 but looking at the histogram we see that 72 is not a typical wait time. It would be misleading to tell visitors they should expect to wait 72 minutes. Following a short eruption one should expect to wait approximately 57 minutes and following a long eruption one should expect to wait approximately 81 minutes before the next eruption. These two values are very near the centers of the two groups of wait times.

Section 6.2 Parameters and Estimators

6.15 σ 6.17 π

6.19 b the mean

6.21

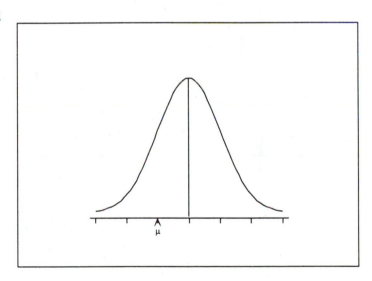

6.23   V will underestimate σ^2. For a sample size of 100, V will estimate 99% of σ^2. For a sample size of 100 the bias is not a serious problem.

6.25   Parameter of interest is the mean time for checking reservations. Possible estimators: $\bar{y}$ or $\bar{y}_T$
$\bar{y} = 42.35$, $\bar{y}_{T.10} = 41.87$

```
Stem-and-leaf of time      N = 20
Leaf Unit = 1.0

     2       2 79
     7       3 01234
     9       3 99
    (2)      4 02
     9       4 578
     6       5 011
     3       5 88
     1       6 3
```

The stem and leaf plot indicates no long tails so $\overline{y}$ = 42.35 is the preferred statistic.

6.27   Because some animals are very slow and may never respond, the distribution is probably skewed right in which case the median is the preferred parameter.

6.29

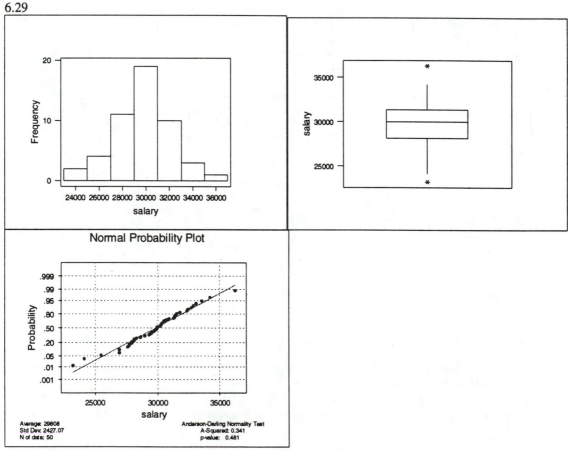

The histogram, boxplot, and normal probability plot all strongly suggest that the parent population is normally distributed. In this case $\overline{y}$ is the best estimator of the mean salary.

Section 6.3 Estimating Certain Population Parameters

6.31   a   The general shape is skewed right.

       b   The boxplot and normal probability plot also indicate that the general shape is skewed right.

       c   The median M = 8 is the estimate of the center of the distribution. The Q-spread = 20 - 5 = 15 is the estimate of the variability.

6.33   The shape of the distribution is skewed left. The median income is the preferred measure of central tendency. It is not possible to calculate the median exactly because the raw data are not given here. It is, however, somewhere between $30,000 and $39,999. We could approximate its value to be around $35,000.

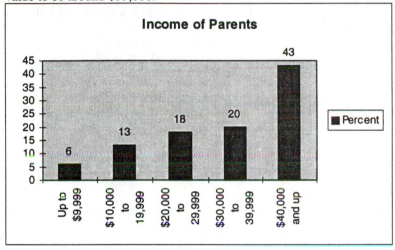

6.35   a   The normal probability plot does not rule out normality.

       b

```
Stem-and-leaf of salinity   N = 48
Leaf Unit = 1.0

    1      3 4
    6      3 56679
   14      4 00002234
  (13)     4 6666777888999
   21      5 00112333
   13      5 68899
    8      6 000123
    2      6 7
    1      7
    1      7 8
```

The stem and leaf plot appears bell shaped with one large value at 78.

c   **Letter Value Display**

	DEPTH	LOWER	UPPER	MID	SPREAD
N=	48				
M	24.5	48.500		48.500	
H	12.5	42.500	57.000	49.750	14.500
	1	34.000	78.000	56.000	44.000

The median and midQ are about the same but the midrange is greater because of the one outlier at 78.

d   Aside from the one observation at 78 the data are very close to being normally distributed. The recommended estimator of the meal salinity level is the sample mean. The estimate is $\overline{y} = 49.54$.

6.37   a   Because the midsummaries are very close in value, we conclude that the distribution is symmetrical.

       b   The boxplot looks very symmetrical, which is consistent with the stem and leaf plot and the midsummary analysis.

       c   The normal probability plot suggests that the distribution is normal.

       d   From parts a-d we conclude that the distribution of the IQ data is close to a normal distribution. We estimate the mean IQ to be $\overline{y} = 101.35$.

6.39   a

```
Stem-and-leaf of yards      N  = 20
Leaf Unit = 1.0

    2    23 89
    4    24 12
    5    24 8
    8    25 114
   10    25 59
   10    26 12
    8    26 558
    5    27 01133
```

Character Boxplot

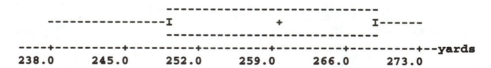

The distribution appears to be short-tailed and slightly skewed left.

       b   The normal probability plot has the general appearance of a short-tailed distribution. This agrees with the assessment given in part a.

       c   The midrange = (238 + 273)/2 = 255.5. This is close in value to the other measures of the center of the distribution.

       d   The distribution is short-tailed; any of the standard measures of the center would suffice as an estimate of the center of the distribution. For simplicity, we use $\overline{y} = 257.85$ as the estimate.

6.41   a   **Stem-and-leaf of score**   **N = 31**
            Leaf Unit = 1.0

```
    1      4 6
    3      5 03
    4      5 5
    6      6 11
   12      6 557889
   (5)     7 00033
   14      7 55567
    9      8 23
    7      8 57
    5      9 002
    2      9 5
    1     10 0
```

        Character Boxplot

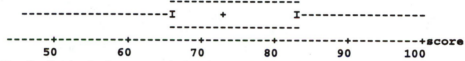

```
 ------+---------+---------+---------+---------+---------+--------+score
      50        60        70        80        90       100
```

        The distribution looks symmetrical and possibly normal.

   b   The normal probability plot does not suggest any departures from normality.

   c   We conclude that the distribution is close to normal and our estimate of the center is $\bar{y} =$
        73.1.

6.43   An estimate of the percent of registered voters in Miami who are Puerto Rican is $p = 45/225 =$
       .20.

Section 6.4   Summary and Review

6.45   a $\bar{y}$

      b $\bar{y}_T$

      c $\bar{y}$

      d   M

      e   M

6.47   A robust estimator

6.49 μ

6.51 π

6.53   a   **Stem-and-leaf of sales   N = 18**
            Leaf Unit = 1000

```
4       0  0001
9       0  23333
9       0  44455
4       0  7
3       0  8
2       1
2       1  2
1       1
1       1
1       1
1       2  0
```

The distribution is skewed right.

b   The midsummaries indicated that the distribution is skewed right. For skewed distributions there is no need to check for normality because the normal distribution is symmetrical.

6.55   **Stem-and-leaf of porosity N = 20**
        Leaf Unit = 1.0

```
 1      1  0
 2      1  3
 5      1  555
 6      1  7
10      1  8889
10      2  01
 8      2  33
 6      2  44
 4      2  677
 1      2  9
```

	DEPTH	LOWER	UPPER	MID	SPREAD
N=	20				
M	10.5		19.500		19.500
H	5.5	16.000	24.000	20.000	8.000
	1	10.000	29.000	19.500	19.000

The midsummaries are close in value so the distribution appears to be symmetrical.

6.57   The normal probability plot shows no departures from normality. This agrees with our assessment in exercises 6.55 and 6.56

6.59   a   **Stem-and-leaf of charge   N = 17**
            Leaf Unit = 1000

```
 1      2  4
 2      2  7
 5      2  999
 8      3  111
(1)     3  2
 8      3  444455
 2      3  6
 1      3  8
```

b   Based on the midsummaries there is very little skewness in the data.

c   Certainly if the distribution is symmetrical there is a chance that it is normally distributed. A boxplot and normal probability plot should be constructed to evaluate further.

6.61   The normal probability plot shows no departures from normality. Based on Exercises 6.59 and 6.60 we should classify the distribution as normal.

6.63   **Character Boxplot**

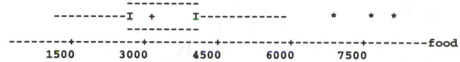

The boxplot and assessment given in Exercise 6.62 leads to the conclusion that the distribution is skewed right. The population median θ is the parameter of interest and its estimate is M = 3165.

6.65   If there are censored data the population median θ, is the best representation of the center of the distribution because it is not affected by the observations that are censored .

6.67   **Stem-and-leaf of salary   N = 48**
Leaf Unit = 1000

```
   1       5 5
   2       6 0
   5       6 589
  12       7 0000034
  20       7 55556678
  23       8 011
  (4)      8 5566
  21       9 0000124
  14       9 59
  12      10 113
   9      10 59
   7      11 02
   5      11 5
   4      12 013
   1      12
   1      13 0
```

Letter Value Display

	DEPTH	LOWER	UPPER	MID	SPREAD
N=	48				
M	24.5	85000.000		85000.000	
H	12.5	74824.500	100443.000	87633.750	25618.500
E	6.5	70000.000	111012.500	90506.250	41012.500
D	3.5	66640.000	120500.000	93570.000	53860.000
C	2.0	60000.000	123300.000	91650.000	63300.000
	1	55502.000	130000.000	92751.000	74498.000

Character Boxplot

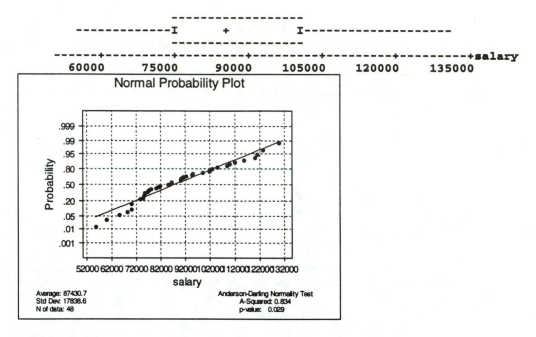

The midsummaries increase and then decrease giving no useful information. The boxplot looks slightly skewed right. The normal probability plot shows substantial departures from normality but no distinct shape. The stem and leaf plot shows that the distribution is multimodal; this explains why the above information is inconclusive. For multimodal distributions it is recommended that we not give a single measure of the center but rather determine what is causing the different modes. If we are successful, then the data are separated into different groups and then summarized. For these data there is insufficient information to determine the cause of the different modes.

6.69 **Character Dotplot**
1 Points missing or out of range

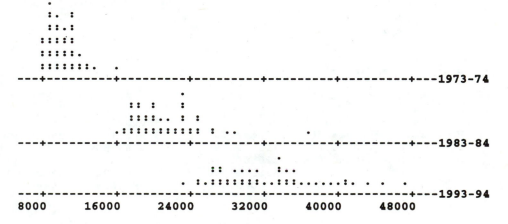

Descriptive Statistics

Variable	N	N*	Mean	Median	TrMean	StDev	SEMean
1973-74	50	1	10113	9891	9998	1585	224
1983-84	51	0	21361	20657	21044	3744	524
1993-94	51	0	32433	31200	32153	5636	789

Variable	Min	Max	Q1	Q3
1973-74	7604	15667	8929	11121
1983-84	15895	36564	18505	23044
1993-94	23300	47000	27600	35200

The side-by-side dotplots and the descriptive statistics show that the mean salary has increased about $10,000 each 10-year period and the standard deviation increased some $2,000 each 10-year period. So not only are salaries increasing they are becoming more variable.

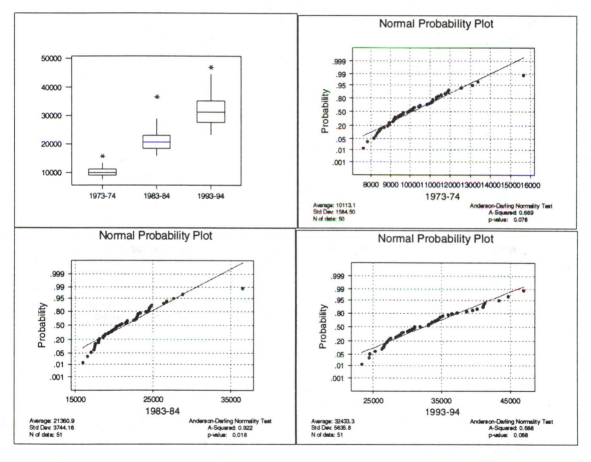

The boxplots and normal probability plots show that all three distributions are skewed right To evaluate the difference we should compare population medians.

6.71   (The graphs of your simulated data should be similar to the graphs given here.)

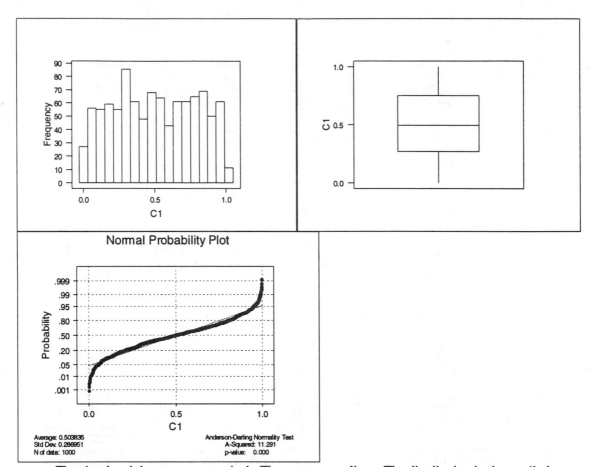

The simulated data are symmetrical. There are no outliers. The distribution is short-tailed.

Descriptive Statistics

Variable	N	Mean	Median	TrMean	StDev	SEMean
C1	1000	0.50384	0.49614	0.50389	0.28695	0.00907

Variable	Min	Max	Q1	Q3
C1	0.00102	0.99982	0.26880	0.75303

Because of the symmetry, the mean, median and trimmed mean are all very close to .5, the midpoint of the interval from 0 to 1. Had we generated simulated values between 1 and 5 the mean should be close to 3.

Descriptive Statistics

Variable	N	Mean	Median	TrMean	StDev	SEMean
C2	1000	3.0344	3.0600	3.0372	1.1408	0.0361

Variable	Min	Max	Q1	Q3
C2	1.0085	4.9948	2.0151	4.0095

246

The simulated data between 1 and 5 has a mean, median and trimmed mean very close to 3, the midpoint between 1 and 5. A histogram, boxplot and normal probability plot of the data will be very similar to the results obtained above except that the data are between 1 and 5 instead of between 0 and 1.

6.73
```
Stem-and-leaf of total94    N  = 51
Leaf Unit = 10

    1      8 3
    3      8 44
    7      8 6777
   18      8 88888999999
   22      9 0011
   (6)     9 222233
   23      9
   23      9 666
   20      9 889
   17     10 0011111
   10     10 2223
    6     10 4455
    2     10 6
    1     10 8
```
The stem and leaf plot suggest that the data are bimodal.

The following dotplot shows two distinct groups, those states where fewer than 40%(group 1) took the SAT and those states where more than 40% took the SAT (group 2).

Character Dotplot

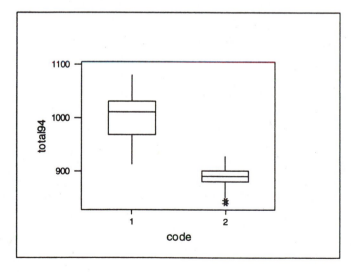

The side-by-side boxplots again show the separation into two distinct groups. Both groups are reasonably symmetrical without many outliers, therefore the mean would suffice as a measure of the center of each distribution.

```
Descriptive Statistics

Variable      code        N      Mean    Median   TrMean     StDev    SEMean
total94         1        27    1004.0    1011.0   1004.6      43.8       8.4
                2        24    887.83    890.00   888.32     23.38      4.77

Variable      code      Min       Max        Q1        Q3
total94         1      913.0    1080.0     969.0    1031.0
                2      838.00    927.00    879.00    900.50
```

The descriptive statistics show that the mean SAT for group 1 is 1004 and the mean SAT for group 2 is 887.83, a clear difference. The standard deviations show that the scores for group 1 are more variable than those for group 2.

The following scatterplot of SAT scores versus the percent taking the exam shows a clear downward trend (Correlation of total94 and percent = -0.875). The higher the percent taking the exam, the lower the scores. This is explained by the fact that in those states where the percent taking the exam is low only the better students are taking the exam. They are the students that are transferring out of state to go to more prestigious schools.

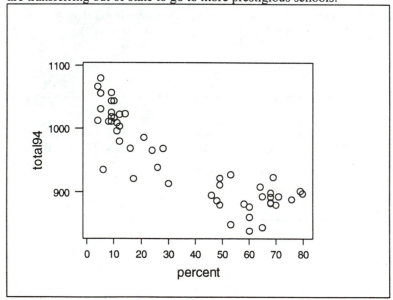

248

Unit 2 [Chapters 4-6]

Significant Ideas and Review Exercises

U2.1 a   The parameter of interest is the Bernoulli proportion π.

     b   It would be impossible to conduct a census, we should take a random sample.

     c   With a sample size of 500 we should be within $\pm.045$ of the true proportion. The sample size is adequate.

U2.3 a   In the Christmas tree example π represents the proportion that will buy an artificial tree.

     b   The ratio 135/500 is from the sample, therefore it is one of the possible values of p. Another sample of 500 might yield a different value of p.

     c   We cannot say that 27% of all buyers will purchase artificial trees. We can only say that our estimate of the number of buyers that will purchase artificial trees is 27%.

U2.5 a   S = {www, wwm, wmw, mww, wmm, mwm, mmw, mmm}

     b $P(www) = (.4)^3$, $P(wwm) = (.4)^2(.6)^1$, $P(wmw) = (.4)^2(.6)^1$, $P(mww) = (.4)^2(.6)^1$, $P(wmm) = (.4)^1(.6)^2$, $P(mwm) = (.4)^1(.6)^2$, $P(mmw) = (.4)^1(.6)^2$, $P(mmm) = (.6)^3$

     c   A = {wwm, wmw, mww}

     d $P(A) = 3(.4)^2(.6)^1 = .288$

U2.7 a   31,892/116,564 = 27%

     b   (6680+1561+434)/31892 = 27%

U2.9   S={(L1,L1,L1),(L1,L1,L2),...,(L3,L3,L3)}

       S has 27 equally likely outcomes, so the probability of any outcome is 1/27.

U2.11 Number each president's picture from 1 to 40. S = { (p_1,p_2,p_3,p_4,p_5)| p_i = 1,2,...,40; i = 1,2,3,4,5 } The number of outcomes in S is 40^5 and each of these is equally likely. The event A that all 5 pictures are different requires that $p_1 \neq p_2 \neq ... \neq p_5$. There would be 40 choices for p_1, 39 choices for p_2, and so on. Thus A has 40*39*38*37*36 elements and P(A) = (40*39*38*37*36)/40^5 = .771

U2.13 Given P(HB) = .3, P(OW) = .4, and P(HB and OW) = .1

     a   n = 15,π = .3; Using Binomial probability tables, $P(Y \leq 3) = .005 + .031 + .092 + .170 = .298$

     b   n = 15, π = .4; Using Binomial probability tables, $P(Y \leq 3) = 0 + .005 + .022 + .063 = .090$

     c   n = 15, π = .10; Using Binomial probability tables, $P(Y \leq 3) = .206 + .343 + .267 + .129 = .945$

     d   E(HB) = 15(.3) = 4.5

     e   E(OW) = 15(.4) = 6

f   E(HB and OB) = 15(.1) = 1.5

U2.15 Let Y be the number on welfare out of 8 families. Y is binomial with n = 8 and π = .7.
P(Y = 8) = .058

U2.17 Let X be the number of shots until he misses. Because he may never miss, the range of X is
{1,2,3,4,---}. X is discrete.

U2.19 Let Y be the distance jumped. Y is normal with μ = 22, σ = 3, Z =(Y - 22)/3 and Y = 22 + Z(3).
   a   P(Y > 24) = P(Z > .67) = 1 - .7486 = .2514
   b   The 30th percentile is z = -.52 so y = 22 + (-.52)(3) = 20.44
   c   Z = (23 - 22)/(3/$\sqrt{32}$) = 1.89   P($\bar{y}$ > 23) = 1 - .9706 = .0294

U2.21 Y = 20 + Z(5). The top 25% is the 75th percentile which is z = .67. So y = 20 + .67(5) = 23.35

U2.23 P(X > 13) = P(Z > (13-10.4)/2.5) = P(Z > 1.04) = 1 - .8508 = .1492

U2.25 a   U2.27 d   U2.29 d

U2.31 a $(2/3)^3$ = .296
   b $(1/3)^3 + 3(1/3)^2(2/3)^1$ = .259
   c $(1/3)^3$ = .037

U2.33 a   z = (15 - 12)/1.5 = 2   P(X > 15) = 1 - .9772 = .0228
   b   z = (8 - 12)/1.5 = -2.67   z = (11 - 12)/1.5 = -.67 P(8 < X < 11) = .2514 - .0038 = .2476
   c   x = 12 + 1.28(1.5) = 13.92

U2.35 a   **Stem-and-leaf of trials   N = 25**
       Leaf Unit = 1.0

```
        3       0 334
        9       0 667899
       (5)      1 01244
       11       1 5568
        7       2 013
        4       2 56
        2       3 1
        1       3
        1       4
        1       4 8
```

   b   **Letter Value Display**

	DEPTH	LOWER	UPPER	MID	SPREAD
N=	25				
M	13.0		14.000	14.000	
H	7.0	8.000	20.000	14.000	12.000
	1	3.000	48.000	25.500	45.000

   c   The midsummaries indicate right skewness.
   d $\bar{y}$ = 14.96 s = 10.22

e   **Character Boxplot**

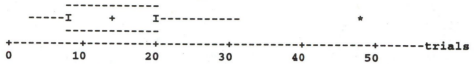

```
                    -------------
            -----I      +       I-----------                      *
                    -------------
          +---------+---------+---------+---------+---------+------trials
          0        10        20        30        40        50
```

From the shape of the boxplot we classify the distribution as skewed right.

f   Parameters that measure the center and the variability are the median and the q-spread. Estimates are M = 14 and Q-spread = 20 - 8 = 12 respectively.

U2.37
```
MTB > Random 60 c1;
SUBC>    Integer 1 6.
MTB > Tally C1;
SUBC>    Counts;
SUBC>    Percents.
```

Summary Statistics for Discrete Variables

C1	Count	Percent
1	10	16.67
2	14	23.33
3	10	16.67
4	11	18.33
5	8	13.33
6	7	11.67
N=	60	

Out of 60 rolls we expect each value to appear 10 times. From the simulation we see that a 2 came up 14 times and the 6 came up only 7 times. A second simulation should give similar results.

Following is the result of five more simulations:

Summary Statistics for Discrete Variables

C1	Count	Percent		C2	Count	Percent		C3	Count	Percent
1	10	16.67		1	10	16.67		1	12	20.00
2	15	25.00		2	9	15.00		2	10	16.67
3	9	15.00		3	8	13.33		3	11	18.33
4	6	10.00		4	11	18.33		4	9	15.00
5	6	10.00		5	12	20.00		5	12	20.00
6	14	23.33		6	10	16.67		6	6	10.00
N=	60			N=	60			N=	60	

C4	Count	Percent		C5	Count	Percent
1	7	11.67		1	15	25.00
2	9	15.00		2	10	16.67
3	9	15.00		3	7	11.67
4	13	21.67		4	7	11.67
5	11	18.33		5	9	15.00
6	11	18.33		6	12	20.00
N=	60			N=	60	

U2.39
```
MTB > Random 50 c1;
SUBC>    Bernoulli .55.
MTB > Tally  C1;
SUBC>    Counts;
SUBC>    Percents.

Summary Statistics for Discrete Variables
C1  Count Percent
 0    21    42.00
 1    29    58.00
N=    50
```
From the simulation we find that 29 out of 50 (58%) graduated. This is very close to the population proportion of 55%. A second simulation should yield similar results.

U2.41
```
MTB > Set c1
DATA>    1( 0 : 30 / 1 )1
DATA>    End.
MTB > PDF c1 c2;
SUBC>    Binomial 30 .13.
x   p(x)
0   0.015331
1   0.068726
2   0.148907
3   0.207671
4   0.209461
5   0.162754
6   0.101331
7   0.051914
8   0.022302
9   0.008146
10  0.002556
11  0.000694
12  0.000164
13  0.000034
14  0.000006
15  0.000001
16  0.000000
17  0.000000
18  0.000000
19  0.000000
20  0.000000
21  0.000000
22  0.000000
23  0.000000
24  0.000000
25  0.000000
26  0.000000
27  0.000000
28  0.000000
29  0.000000
30  0.000000
```
The probability that none of the 30 go into internal medicine is $p(0) = 0.015331$. The number expected is $30(.13) = 3.9$. The standard deviation is $\sqrt{30(.13)(.87)} = 1.842$.

U2.43
```
MTB > Random 50 c1;
SUBC>    Normal 70.0 8.0.
MTB > Histogram C1;
SUBC>    MidPoint;
SUBC>    Bar.
```

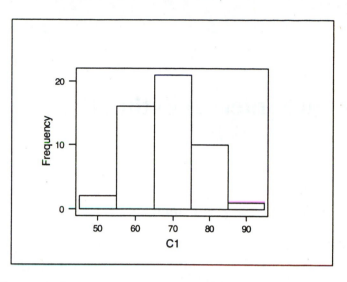

U2.45 **Stem-and-leaf of rate   N = 51**
Leaf Unit = 0.10

```
     4      1 5555
    15      1 66666677777
   (18)     1 888888888899999999
    18      2 0000011111
     8      2 2223
     4      2 445
     1      2
     1      2
     1      3
     1        3 2
```

Letter Value Display

	DEPTH	LOWER	UPPER	MID	SPREAD
N=	51				
M	26.0		1.900	1.900	
H	13.5	1.700	2.050	1.875	0.350
E	7.0	1.600	2.200	1.900	0.600
D	4.0	1.500	2.400	1.950	0.900
C	2.5	1.500	2.450	1.975	0.950
B	1.5	1.500	2.850	2.175	1.350
	1	1.500	3.200	2.350	1.700

The midsummaries get progressively larger therefore the distribution is skewed right.

7

Confidence Interval Estimation

Section 7.1 Confidence Interval for a Population Proportion

7.1   The length of the interval will increase.

7.3 $p = 722/1183 = .61$
.61 ± 2.58$\sqrt{(.61)(.39)/1183}$ The confidence interval is .573 to .647
A 95% confidence interval would be narrower because $z^* = 1.96$ instead of 2.58.

7.5 $p = 180/900 = .2$
.2 ± 1.645$\sqrt{(.2)(.8)/900}$ The confidence interval is .178 to .222
$.03 = 1.645\sqrt{(.2)(.8)}/\sqrt{n}$, so n = 482

7.7 $p = 880/1000 = .88$
.88 ± 1.645$\sqrt{(.88)(.12)/1000}$ The confidence interval is .863 to .897

7.9 $p = 1440/2000 = .72$
.72 ± 2.33$\sqrt{(.72)(.28)/2000}$ The confidence interval is .697 to .743
No, either the true proportion is inside the interval or it is not. Probability is only relevant prior to sampling.

7.11 $p = 708/1200 = .59$
.59 ± 1.645$\sqrt{(.59)(.41)/1200}$ The confidence interval is .567 to .613
Maximum margin of error is 1.645$\sqrt{(.5)(.5)/1200} = .024$
$.03 = 1.645\sqrt{(.5)(.5)}/\sqrt{n}$, so n = 752

7.13 $.03 = 2.33\sqrt{(.6)(.4)}/\sqrt{n}$, so n = 1448

7.15   If we add and subtract the margin of error (sampling error) of ±4.5% to either of the percents, 36% or 33%, we obtain an interval that contains the other percent. In other words, each sample percent is a possible candidate for the other population percent. Statistically there is no difference between the two population percents. The sampling error of ±4.5% is the maximum margin of error for a 95% confidence interval based on a sample of size 500. By our calculations we find $1.96\sqrt{(.5)(.5)/500} = .044 \approx 4.5\%$. A confidence interval for the percent that thought that federal involvement would make the system worse is 36% ±4.5% or 31.5% to 40.5%. A confidence interval for the percent that thought that federal involvement saw improvement is 33% ±4.5% or 28.5% to 37.5%.

Section 7.2 Confidence Interval for μ Based on $\overline{y}$: The z-Interval

7.19   a   465.8 ± 1.96*50/√60 The confidence interval is 453.15 to 478.45
       b   465.8 ± 2.58*50/√60 The confidence interval is 449.15 to 482.45

7.21   Interval A is the 95% interval because a 95% confidence interval is generally shorter than a 99% confidence interval.

7.23   1325 ±1.96*42.4/√100 The confidence interval is 1316.69 to 1333.31. We cannot say that there is a 95% chance that the mean living area is inside the interval. Either the mean is inside the interval or it is not. Probability (95% chance) is only relevant prior to sampling.

7.25   6.5 ± 1.96*2.3/√22 The confidence interval is 5.54 to 7.46

7.27   5.2 ± 2.58*(.24)/√40 The confidence interval is 5.1 to 5.3
       If the students had taken a random sample of 100 cups the resulting interval would be narrower because the margin of error would have been divided by √100 instead of √40.

7.29   6.1 ± 2.33*2.1/√100 The confidence interval is 5.61 to 6.59
       .3 = 2.33*2.1/√n so n = 267

7.31   78460 ± 1.645*(22000)/√40 The confidence interval is 72737.86 to 84182.14
       The Central Limit Theorem says that the z-interval can be applied to non-normal populations if the sample size is sufficiently large. In the case of severe skewness or symmetric long tails, a sample size of 40 may not be adequate. Therefore, it is a good idea to always check for skewness and long tails.

Section 7.3 Confidence Interval for μ Based on $\overline{y}$: The t-Interval

7.33   69.2 ± 2.776*3.27/√5 The confidence interval is 65.1 to 73.3

7.35   The midsummaries are close in value suggesting that the distribution is symmetric. With a sample size of 15 it is a good idea to check for normality. This is accomplished with a normal probability plot. Assuming the conditions for a t-interval we have 57.17 ± 2.624*8.21/√15. The confidence interval is 51.6 to 62.7

7.37   250.03 ± 2.000*41.44/√62 The confidence interval is 239.50 to 260.56

7.39 With a sample size of 48 the normality condition is not so crucial because of the Central Limit Theorem. We should, however, still be concerned with long tails and severe skewness. To find the t-interval we have $49.54 \pm 2.685*9.27/\sqrt{48}$ The confidence interval is 45.95 to 53.13

7.41 The placebo group appears normally distributed but the treatment group appears skewed. By removing the one extreme outlier the data appear more normally distributed. Because the researcher stated that outliers of this type were typical we probably should not remove the outlier.

7.43 Based on the normal probability plot these data are severely skewed right. Because of the skewness, the median is the recommended parameter to measure a typical return on an investment. For a t-interval the data should be more normally distributed. Clearly this is not the case.

7.45 a The boxplot and normal probability plot indicate strong right skewness.
      b The sample size is 629 with 90 missing observations. With such a large sample size it is permissible to find a confidence interval for the mean ozone using a t-interval.
      c $23.693 \pm 2.58*19.276/\sqrt{629}$ The confidence interval is 21.71 to 25.68.
      d With such a large sample size the validity is not affected by the shape of the population distribution.

Section 7.4 Confidence Interval for the Population Median

7.47 a 6 b 5 c 4

7.49 n = 12, Location of C = 4; $C_L = 20$, $C_H = 45$

7.51 a
```
     Stem-and-leaf of 34kV        N  = 19
     Leaf Unit = 1.0

       (13)   0  0001234446788
         6    1  2
         5    2
         5    3  1236
         1    4
         1    5
         1    6
         1    7  2
```

Letter Value Display

	DEPTH	LOWER	UPPER	MID	SPREAD
N=	19				
M	10.0	6.500		6.500	
H	5.5	2.970	21.905	12.437	18.935
E	3.0	0.960	33.910	17.435	32.950
D	2.0	0.780	36.710	18.745	35.930
	1	0.190	72.890	36.540	72.700

The midsummaries increase and the stem and leaf plot shows severe right skewness.

    b $n = 19$, Location of C = 5; $C_L = 2.78$, $C_H = 31.75$

7.53   a $(25 - 1.96*\sqrt{25})/2 = 8$
          b $(50 - 1.96*\sqrt{50})/2 = 19$
          c $(100 - 1.96*\sqrt{100})/2 = 41$
          d $(200 - 1.96*\sqrt{200})/2 = 87$

7.55   a   Location of C = $(25 - 1.96*\sqrt{25})/2 \approx 8$
               (27,45)
          b   Location of C = $(25 - 1.96*\sqrt{25})/2 \approx 8$
               (15,36)
          c   Location of C = $(24 - 1.96*\sqrt{24})/2 = 7.2 \approx 8$
               (12,36)

7.57   Location of C = $(26 - 1.96*\sqrt{26})/2 = 8$
       (75,500, 100,000)

7.59   The normal probability plot indicates that the data are skewed.
       $n = 14$, Location of C = 3; $C_L = 16$, $C_H = 24$

7.61   a   Small mesh: Location of median $(739 + 1)/2 = 370$ M = 33
               Large mesh: Location of median $(767 + 1)/2 = 384$ M = 34
          b   Small mesh: Location of quartile $(370 + 1)/2 = 185.5$ $Q_1 = 31$ $Q_3 = 35$
               Large mesh: Location of quartile $(384 + 1)/2 = 192.5$ $Q_1 = 33$ $Q_3 = 36$
          c   It seems that the length of fish caught with the large mesh codend is slightly larger than
               with the small mesh codend.
          d   Small mesh: Location of C = $(739 - 2.58*\sqrt{739})/2 = 335$ (33, 34)
               Large mesh: Location of C = $(767 - 2.58*\sqrt{767})/2 = 348$ (34, 35)
               There is no overlap in the two intervals. The length of fish caught with a small mesh
               codend is somewhere between 33 and 34 cm and with the large mesh codend is somewhere
               between 34 and 35 cm.

Section 7.5 Robust Confidence Interval for the Center of a Distribution

7.63   The boxplot shows a symmetric long-tailed distribution.
       $34.2 \pm 2.33*(8.373)/\sqrt{40}$ The confidence interval is (31.12, 37.28)

7.65
```
Stem-and-leaf of time      N  = 50
Leaf Unit = 0.10

        3      3 134
        6      3 568
        8      4 01
       21      4 5566667788899
       (8)     5 01223344
       21      5 66677889999
       10      6 344
        7      6 78
        5      7 14
        3      7
        3      8 34
        1      8 8
```
The distribution appears long-tailed and therefore the confidence interval will be based on $\bar{y}_T$.
We will trim 10%. $\bar{y}_{T\ .10} = 5.27$ $s_W = .939$, $s_T = 1.053$
$5.27 \pm 1.96*(1.053)/\sqrt{40}$ The confidence interval is (4.94, 5.60)

7.67
```
Stem-and-leaf of cost      N  = 25
Leaf Unit = 10

        1      3 1
        4      4 457
        8      5 1478
       (6)     6 114688
       11      7 012458
        5      8 4
        4      9
        4     10 0
        3     11 0
        2     12 8
        1     13 5
```
Location of C = (25 - 2.58*$\sqrt{25}$)/2 ≈ 7 The confidence interval is (575,758)

7.69 $153 \pm 1.96*(21.5)/\sqrt{28}$ The confidence interval is (145.04, 160.96)

7.71 $1530 \pm 2.33*285/\sqrt{50}$ The confidence interval is (1436.09, 1623.91)

Section 7.6 Summary and Review

7.73   a   1.645   b   2.33   c   1.753   d   2.602

7.75   a   skewed right   b   median
        c   Location of C = (25 - 2.58$\sqrt{25}$)/2 = 7; The confidence interval is (18, 53)

7.77   a $.03 = 2.58*\sqrt{(.5)(.5)}/\sqrt{n}$, so n = 1849
        b   p = 1120/1600= .7, $.7 \pm 2.58\sqrt{(.7)(.3)/1600}$ The confidence interval is (.67, .73)

We are very confident that the true percent of convicted felons with a history of juvenile delinquency is somewhere between 67% and 73%.

7.79 a
```
Stem-and-leaf of fee       N = 25
Leaf Unit = 1.0

    2      0 55
    5      1 000
    8      1 555
   (7)     2 0000000
   10      2 5555
    6      3 0000
    2      3 55
```

b From the stem and leaf plot it appears that the data have not violated the normality assumption.

c **Confidence Intervals**

Variable	N	Mean	StDev	SE Mean	90.0 % C.I.	
fee	25	20.60	8.58	1.72	(17.66,	23.54)

7.81 a $p = 960/2400 = .4$, $.4 \pm 2.58\sqrt{(.4)(.6)/2400}$ The confidence interval is $(.374, .426)$

b $.02 = 2.58\sqrt{(.4)(.6)}/\sqrt{n}$, so $n = 3994$

7.83 $6.7 \pm 2.05*(.2)/\sqrt{100}$ The confidence interval is $(6.66, 6.74)$

7.85 Editor $p = 144/1023 = .141$, $.141 \pm 2.58\sqrt{(.141)(.859)/1023}$ $(.113, .169)$
Assoc Ed $p = 49/197 = .249$, $.249 \pm 2.58\sqrt{(.249)(.751)/197}$ $(.17, .328)$
News Ed $p = 278/1132 = .246$, $.246 \pm 2.58\sqrt{(.246)(.754)/1132}$ $(.213, .279)$

All editor $p = 873/4501 = .194$, $.194 \pm 2.58\sqrt{(.194)(.806)/4501}$ $(.179, .209)$

```
Editor   x------------x
Assoc Ed              x-----------------------------x
News Ed                     x------------x

All Ed             x----x
------|---------|---------|---------|---------|---------|----
    .1        .15        .2        .25        .3        .35
```

It is easy to see that the precision of an interval is dependent on the sample size. The interval for Associate Editors has the least precision and the interval for all key editor positions has the most precision. There is no overlap between the first and third intervals whereas the third interval is completely contained in the second interval. The all key editor positions interval is sort of an average of the three previous intervals. We are confident that the percent of women holding key editor positions is somewhere between 18% and 21%.

7.87 Location of $C = (51 - 2.33*\sqrt{51})/2 \approx 18$. The confidence interval is $(3, 12)$

7.89 Waitime 1 $57.109 \pm 1.9822*8.148/\sqrt{110}$ The confidence interval is $(55.569, 58.649)$
Waitime 2 $81.164 \pm 1.9727*7.303/\sqrt{189}$ The confidence interval is $(80.116, 82.212)$

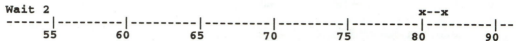

The precision of the waitime 2 interval is greater (narrower interval) because it is based on a larger simple size. Also the standard deviation of the waittime 2 data is slightly smaller. This causes the interval to be shorter but the real difference in precision of the two intervals is because of the sample sizes. There is a clear distinction between the wait times for the two types of eruptions.

7.91 Estimate σ with Range/4 = 20/4 = 5.
      .5 = 1.96*5/$\sqrt{n}$, so n = 385

7.93 To construct a t-interval when the sample size is 16, it is important that the reading speed of 5th graders be approximately normally distributed.
      285 $\pm$ 2.131*48/$\sqrt{16}$ The confidence interval is 259.43 to 310.57

7.95 82.4 $\pm$ 1.691*23.5/$\sqrt{35}$ The confidence interval is 75.683 to 89.117

7.97 2 = 15/$\sqrt{n}$, so n = 57

7.99 p = 10540/17000 = .62
      .62 $\pm$2.58$\sqrt{(.62)(.38)/17000}$ The confidence interval is .61 to .63

7.101 a   42,300 $\pm$ 2.602*4000/$\sqrt{16}$ The confidence interval is 39698 to 44902
      b   With a sample size of 16 it is important that the distribution of incomes for new Ph.D.'s in psychology be approximately normal.
      c   By increasing the sample size.

7.103 38,394 $\pm$ 2.68*8482/$\sqrt{49}$ The confidence interval is 35146.61 to 41641.39

7.105 a

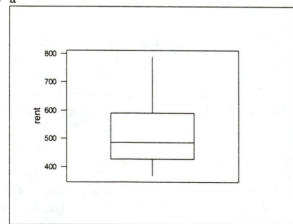

The data are moderately skewed right.

b   With a sample size of 46 and the fact that the skewness is not severe, the normality assumption is not crucial.

c   **Confidence Intervals**

Variable	N	Mean	StDev	SE Mean	99.0 % C.I.	
rent	46	514.9	115.0	17.0	(469.3,	560.5)

7.107

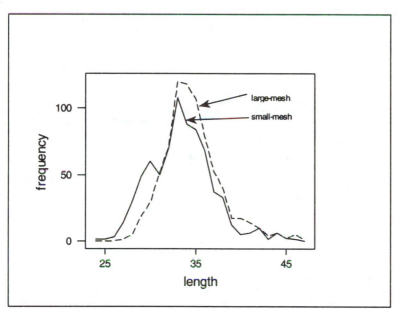

The two distributions do not deviate significantly from normality and the sample sizes are large enough that normality is really not an issue. Confidence intervals for the means are certainly appropriate.

$\bar{y}_{small} = 24700/739 = 33.424$ $s_{small} = \sqrt{[1610516 - (24700)^2/739]/738} = 32.613$

$\bar{y}_{large} = 26550/767 = 34.615$ $s_{large} = \sqrt{[2046288 - (26550)^2/767]/766} = 38.362$

Small mesh confidence interval: $33.424 \pm 2.58*32.613/\sqrt{739}$ (30.329, 36.519)

Large mesh confidence interval: $34.615 \pm 2.58*38.362/\sqrt{767}$ (31.041, 38.189)

In comparison the small mesh confidence interval for the median is (33, 34) and the large mesh confidence interval for the median is (34, 35). Both intervals for the mean are centered at approximately the same place as the intervals for the medians. The median intervals are much more precise than the mean intervals. In general, it appears that the large mesh codend yields catches 1 cm larger than small mesh codend.

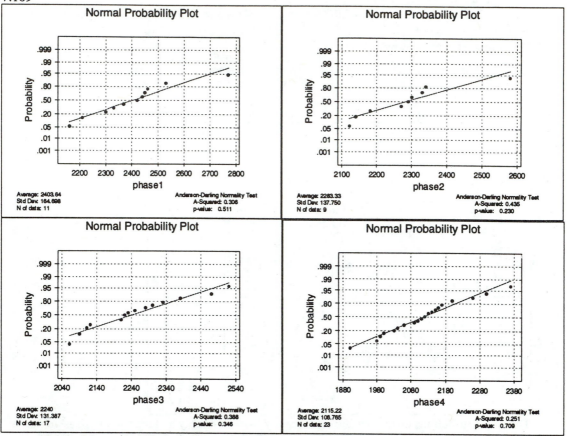

None of the probability plots show a significant departure from normality.
Confidence Intervals

Variable	N	Mean	StDev	SE Mean	95.0 % C.I.	
phase1	11	2403.6	164.7	49.7	(2293.0,	2514.3)
phase2	9	2283.3	137.7	45.9	(2177.4,	2389.2)
phase3	17	2240.0	131.4	31.9	(2172.4,	2307.6)
phase4	23	2115.2	106.8	22.3	(2069.0,	2161.4)

```
phase 1                                    x--------------------x
phase 2                     x---------------------x
phase 3                     x-------------x
phase 4        x--------x
        ------|---------|---------|---------|---------|---------|----
            2000      2100      2200      2300      2400      2500
```

Phases 1, 3 and 4 appear to be non-overlapping periods of time. Phase 2, however, appears to be a mixture of phases 1 and 3.

7.111 a

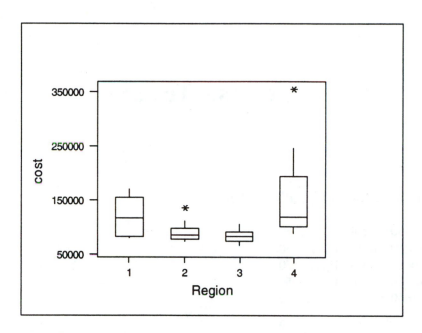

b   The cost of homes in region 4 are more variable than in any other region. There is very little variability in region 3. The median cost of homes in regions 1 and 4 are about the same but are substantially higher than in regions 2 and 3. Because of the skewness in region 4 the mean cost will be much higher than the median cost. The mean and median will be about the same in region 3.

c   **Descriptive Statistics**

Variable	Region	N	Mean	Median	TrMean	StDev	SEMean
1994	1	11	123536	116800	123144	33385	10066
	2	17	90335	84900	88493	16211	3932
	3	23	82791	82900	82524	10736	2239
	4	14	154750	119250	143525	75794	20257

Variable	Region	Min	Max	Q1	Q3
1994	1	80000	170600	82400	154900
	2	72800	135500	77400	97100
	3	66200	105000	73600	91300
	4	89200	355000	100525	193750

   Confidence Intervals

Variable	N	Mean	StDev	SE Mean	98.0 % C.I.	
Region1	11	123536	33385	10066	(95710,	151363)
Region2	17	90335	16211	3932	(80175,	100495)
Region3	23	82791	10736	2239	(77175,	88408)
Region4	14	154750	75794	20257	(101051,	208449)

8

Hypothesis Testing

Section 8.1 Introduction to Hypothesis Testing

8.1   a H_0: $\mu = 50$ versus H_a: $\mu > 50$.
       b H_0: $\mu = 50$ versus H_a: $\mu < 50$.
       c H_0: $\mu = 50$ versus H_a: $\mu > 50$.
       d H_0: $\mu = 50$ versus H_a: $\mu < 50$.
       e H_0: $\mu = 50$ versus H_a: $\mu \neq 50$.

8.3
a   Left-tailed   b   Right-tailed   c   Left-tailed

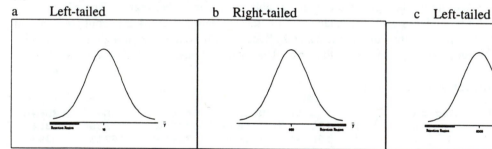

8.5 H_0: $\pi = .10$ versus H_a: $\pi > .10$.

8.7   c

8.9   No, there is the possibility of a Type II error.

8.11   No, Type I and II errors are not complementary events.

8.13   Both are wrong. A p-value is the probability that the null hypothesis is rejected **when** it is actually true.

8.15   Type I : concluding that the fire is not out and it actually is.
       Type II : concluding that the fire is out and it actually is not out.
       Committing a Type II error is more serious.

8.17 H_0: $\mu = .18$ versus H_a: $\mu > .18$

A type II error is committed if the geologist concludes that the mean porosity measurement does not exceed 18% when in fact it does.

8.19   a H_0: $\mu = 40$ versus H_a: $\mu \neq 40$
        b   The population standard deviation was given, $\sigma = 5$.
        c   The sample size was 37
        d $\bar{y} = 38.378$ which is 1.97 standard errors below the mean.
        e   p-value = .049 which should reject the null hypothesis
        f   Based on this random sample of size 37, there is a significant difference between the sample mean and the hypothesized population mean.

8.21   If we set up hypotheses as follows: H_0: $\mu = 78$ versus H_a: $\mu > 78$ we find z = (81.4 - 78)/(7/$\sqrt{35}$) = 2.87 which gives p-value = .0021. Based on these results there is highly significant evidence that his class is superior. Because of the sample size of 35 the normality assumption is not that crucial. We should, however, look for severe skewness or extremely long tails.

Section 8.2 Testing a Population Proportion

8.23 z_{obs} = (.1125 - .1)/($\sqrt{(.1)(.9)/1600}$) = 1.67   p-value = .0475
        Based on a random sample of 1600 adults there is significant evidence that more than 10% of the adult population is illiterate.

8.25 z_{obs} = (.55 - .6)/($\sqrt{(.6)(.4)/200}$) = -1.44   p-value = .0749
        Based on a random sample of 200 adults there is moderately significant evidence that less than 60% of the adult population believes that there is too much violence on television.

8.27   .01 = 1.96*($\sqrt{(.25)(.75)/n}$), so n = 7203.

8.29 H_0: $\pi = .33$ versus H_a: $\pi > .33$
        z_{obs} = (.35 - .33)/($\sqrt{(.33)(.67)/600}$) = 1.04   p-value = .1492
        Based on these 600 cases there is insignificant evidence that Cook County exceeds the national average in the number of automobile accident suits.

8.31   a H_0: $\pi \geq .50$ versus H_a: $\pi < .50$
        b z_{obs} = (.46 - .50)/($\sqrt{(.5)(.5)/1022}$) = -2.56
        c   If the z statistic is less than 1.28 in absolute value then the results are insignificant.
        d   The value of z_{obs} is clearly in the rejection region.
        e   The p-value = .0052 is less than the 10% level of significance.
        f   The null hypothesis should be rejected.
        g   Based on this random sample of 1022 adults there is highly significant evidence that the president does not have the support of the majority on this issue.

8.33 H_0: $\pi = .8$ versus H_a: $\pi \neq .8$
$z_{obs} = (.82 - .8)/(\sqrt{(.8)(.2)/200}) = .71$ p-value = 2(.2389) = .4778
Based on a random sample of 200 people there is insufficient evidence to refute the psychologist's claim. For a two-tailed test the value of z_{obs} must be greater than 1.96 in absolute value for the results to be significant at the 5% level of significance. This means that $1.96 = (p - .8)/(\sqrt{(.8)(.2)/200})$. Solving we find p = .855, which is 171 out of 200.

8.35 H_0: $\pi = 1/3$ versus H_a: $\pi < 1/3$
$z_{obs} = (.24 - .33)/(\sqrt{(.33)(.67)/200}) = -2.71$ p-value = .0034
Based on this sample of 200 prisoners there is highly significant evidence that less than 1/3 are returning to jail within 3 years. These results are significant at any α greater than .34%

Section 8.3 Testing a Population Mean

8.37 a   This is a left-tailed test. To conduct a t-test when the sample size is 25 the population distribution should be reasonably symmetrical without any outliers. To complete the test we have $t = (2.33 - 2.5)/(.67/\sqrt{25}) = -1.27$ p-value = .1081 Based on this random sample there is insignificant evidence that the population mean is less than 2.5.

      b   This is a two-tailed test. To conduct a t-test when the sample size is 70 the population distribution should not be severely skewed or have extreme outliers. To complete the test we have $t = (51.6 - 50)/(5.9/\sqrt{70}) = 2.27$ p-value = 2(.0132) = .0264. Based on this random sample there is significant evidence that the population mean is something other than 50.

      c   This is a right-tailed test. To conduct a t-test when the sample size is 14 the population distribution should be close to normal. To complete the test we have $t = (102.3 - 100)/(2.45/\sqrt{14}) = 3.51$ p-value = .0019. Based on this random sample there is highly significant evidence that the population mean is greater than 100.

8.39 H_0: $\mu \leq 5.5$ versus H_a: $\mu > 5.5$
$t_{obs} = (5.65 - 5.5)/(1.52/\sqrt{40}) = .62$ p-value = .2694
Based on a random sample of 40 cars there is insufficient evidence to deny the manufacturer's claim. The normality assumption is not need here because the sample size is 40. It is important, however, that the population distribution not be severely skewed or have extreme outliers.

8.41 a   With 50 observations there is no reason to base the test on any statistic other than $\bar{y}$.

      b $t_{obs} = (299.96 - 300)/(5.61/\sqrt{50}) = -0.05$ p-value = .96 There is no evidence whatsoever to conclude that the mean price differs from 300 pence.

8.43 H_0: $\mu \leq 120$ versus H_a: $\mu > 120$
$t_{obs} = (125.6 - 120)/(20.3/\sqrt{25}) = 1.379$ p-value = .0903

266

Based on 25 randomly selected families there is moderately significant evidence that mean kilowatt usage exceeds 120 kilowatt hours.

8.45 H_0: $\mu \le 72$ versus H_a: $\mu > 72$
        $t_{obs} = (76 - 72)/(14.4/\sqrt{16}) = 1.111$ p-value = .1420
        Based on the information provided there is insufficient evidence that this is a superior class. Having a normally distributed population insures the accuracy of the t-test. On the other hand, for a sample size of 16 we should insist that the population distribution be at least symmetric without outliers.

8.47 H_0: $\mu \le 78$ versus H_a: $\mu > 78$
        $t_{obs} = (82 - 78)/(7/\sqrt{20}) = 2.556$ p-value = .0097
        Based on the provided data there is highly significant evidence that this is a superior class. Because we are using a t-test and the sample size is 20 we assume that the distribution of scores on this exam is at least symmetric without outliers.

8.49   a   The boxplot and normal probability plot indicate that the parent distribution is close to normal.
        b H_0: $\mu = 8$ versus H_a: $\mu \ne 8$
             $t_{obs} = 1.14$ p-value = .29   Based on the sample of 9 people there is insufficient evidence that the mean number of trials to master a task under the influence of the drug is not 8.

8.51   a   The normal probability plot does not rule out normality.
        b H_0: $\mu \ge 98$ versus H_a: $\mu < 98$
        c   Test of the Mean

             Test of mu = 98.000 vs mu < 98.000

Variable	N	Mean	StDev	SE Mean	T	P-Value
temp	12	93.333	3.393	0.980	-4.76	0.0003

             Based on these 12 readings there is highly significant evidence that the mean inlet oil temperature is less than 98 degrees.

8.53   a H_0: $\mu = 7$ versus H_a: $\mu \ne 7$
        b   This is a two-tailed test.
        c   The evidence is insufficient (p-value = .17) to reject the null hypothesis.
        d   Based on these data we are unable to say that the mean surface salinity level is different from 7 parts per thousand.

8.55 H_0: $\mu = 35,000$ versus H_a: $\mu \ne 35,000$
        $t_{obs} = (32,406 - 35,000)/(3591/\sqrt{17}) = -2.98$ p-value = .0089
        Based on the information provided there is highly significant evidence that the mean charge for a coronary bypass in North Carolina is different from $35,000.

8.57   a   The stem and leaf plot indicates that the data may be bimodal or possibly even trimodal.
        b   Because the sample size is 101 it is possible to ignore the possible multimodality of the data and use $\bar{y}$ as a test statistic.

`Test of mu = 8.000 vs mu < 8.000`

Variable	N	Mean	StDev	SE Mean	T	P-Value
thick	101	7.831	2.587	0.257	-0.66	0.26

There is insufficient evidence that the mean thickness of the varves is less than 8 millimeters.

Section 8.4 Testing a Population Median

8.59 a $z_{obs} = (2*14 - 50)/\sqrt{50} = -3.11$; p-value = .0009
There is highly significant evidence that the median is less than 400.

b $z_{obs} = (2*56 - 100)/\sqrt{100} = 1.2$; p-value = .1151
There is insufficient evidence that the median is more than 12.5.

c $z_{obs} = (2*78 - 200)/\sqrt{200} = -3.11$; p-value = 2(.0009)= .0018
There is highly significant evidence that the median is not 75.

8.61 a When the median is substantially smaller than the mean it is suspected that the distribution is skewed right.

b
```
Stem-and-leaf of cost       N  = 46
Leaf Unit = 1.0             N* =  4

   21      0 111111222333333334444
  (12)     0 556778888899
   13      1 111113
    7      1 67
    5      2 34
    3      2
    3      3 23
    1      3
    1      4
    1      4
    1      5
    1      5 9
```

The stem and leaf plot shows that the distribution is severely skewed right, explaining why the mean cost exceeds the median cost.

c Because of the skewness inferences should be made about the population median, not the mean.

d H_0: $\theta \geq 8$ versus H_a: $\theta < 8$

e Because the p-value = .0586 we reject the null hypothesis and conclude that there is mildly significant evidence that the median cost of operating a state law enforcement agency is less than $8 per resident.

8.63 H_0: $\theta = 100$ versus H_a: $\theta \neq 100$ n = 16, T = 7, p-value = 2*P(T ≤ 7) = .804
There is insufficient evidence to conclude that the median mental age is different from 100.

8.65 `Sign test of median = 19.00 versus  N.E.  19.00`

	N	BELOW	EQUAL	ABOVE	P-VALUE	MEDIAN
age	30	5	12	13	0.0963	19.00

There is mildly significant evidence that the median age is not 19.

8.67 $H_0: \theta \geq 27{,}500$ versus $H_a: \theta < 27{,}500$

`Sign test of median = 27500 versus  L.T.  27500`

	N	BELOW	EQUAL	ABOVE	P-VALUE	MEDIAN
income	25	17	0	8	0.0539	26500

There is significant evidence that the median salary for North Carolina social workers is less than $27,500.

8.69 $H_0: \theta \leq 7.3$ versus $H_a: \theta > 7.3$

`Sign test of median = 7.300 versus  G.T.  7.300`

	N	BELOW	EQUAL	ABOVE	P-VALUE	MEDIAN
thick	37	0	0	37	0.0000	10.70

All 37 observations are greater than 7.3, that is, T = 37. There is highly significant evidence that the median varve thickness exceeds 7.3 mm.

The stem and leaf plot appears to be severely skewed right, however, the boxplot and normal probability only show moderate skewness without any outliers. Testing the population median is appropriate, but testing the population mean would not be out of the question because the skewness is not severe.

8.71 a $H_0: \theta = 7$ versus $H_a: \theta \neq 7$

b   This is a two-tailed test.

c   The standardized test statistic is $z = (2T - n)/\sqrt{n} = 1.04$

d   The p-value = .2955. The null hypothesis should not be rejected.

e   Based on a random sample of 75 water samples there is insignificant evidence that the median pH level is different from 7.

Section 8.5 A Robust Large Sample Test of a Population Mean

8.73 $z_{obs} = (19.4 - 20)/(4.3/\sqrt{96}) = -1.37$; p-value = .0853

There is moderately significant evidence that $\mu < 20$.

8.75 $H_0: \mu = 450$ versus $H_a: \mu > 450$

$z_{obs} = (474.6 - 450)/(127.3/\sqrt{80}) = 1.73$ p-value = .0418

There is sufficient evidence that the mean score is higher than 450.

8.77 H_0: $\mu = 8.50$ versus H_a: $\mu \neq 8.50$
$z_{obs} = (9.23 - 8.50)/(1.69/\sqrt{39}) = 2.70$ p-value $= 2(.0035) = .007$
There is highly significant evidence that the mean cost of an evening meal is not $8.50.

8.79 H_0: $\mu = 120$ versus H_a: $\mu \neq 120$
$z_{obs} = (134.60 - 120)/(37.40/\sqrt{60}) = 3.02$ p-value $= 2(.0013) = .0026$
There is highly significant evidence that the city official's claim is not valid.

8.81 H_0: $\mu \geq 10$ versus H_a: $\mu < 10$
$z_{obs} = (9.6 - 10)/(2.6/\sqrt{60}) = -1.19$ p-value $=.1170$
Based on a random sample of 100 picture tubes there is insignificant evidence that the mean life
is less than 10 years. We cannot refute the manufacturer's claim.

8.83 H_0: $\mu = 95$ versus H_a: $\mu \neq 95$
n = 30, k = 24; $\bar{y}_{T.10} = 98.75$, $s_W = 11.33$ $s_T = 12.72$ $z_{obs} = (98.75 - 95)/(12.72/\sqrt{24}) = 1.44$
p-value $= 2(.0749) = .1498$
Based on a random sample of 30 reading scores there is insufficient evidence that the mean
reading score is different from 95.

8.85   a   The boxplot and the normal probability plot shows a symmetric long-tailed distribution
           which contradicts the conclusion given in Exercise 8.84.
       b   Because of the symmetric long tails $\bar{y}_T$ is the recommended estimator of the parallax of the
           sun.

Section 8.6 Summary and Review

8.87 b   8.89 d   8.91 c   8.93 b

8.95 H_0: $\pi = .70$ versus H_a: $\pi > .70$.

8.97 H_0: $\mu = 600$ versus H_a: $\mu \neq 600$.
$t_{obs} = (642 - 600)/(135/\sqrt{42}) = 2.02$ p-value $= 2(.025) = .05$
There is significant evidence that the mean is not 600.

8.99 H_0: $\mu = 2{,}000$ versus H_a: $\mu > 2{,}000$.
$t_{obs} = (2080 - 2000)/(240/\sqrt{60}) = 2.58$ p-value $=.0062$
There is significant evidence that the county agent is right and his county needs more than 2000
pounds of lime per acre.

8.101 H_0: $\pi = .70$ versus H_a: $\pi \neq .70$.
$z_{obs} = (.77 - .70)/(\sqrt{(.7)(.3)/200}) = 2.16$ p-value $=2(.0154) = .0308$
Based on a random sample of 200 convicted felons there is significant evidence to contradict the
claim that 70% have a history of juvenile delinquency.

8.103 H_0: $\pi = .52$ versus H_a: $\pi < .52$.
$z_{obs} = (.4375 - .52)/(\sqrt{(.52)(.48)/32}) = -.93$ p-value =.1762
Based on a random sample of 32 awarded PhDs there is insignificant evidence that the percent awarded to US students is unusually low.

8.105 H_0: $\mu = 40$ versus H_a: $\mu > 40$.

8.107 H_0: $\mu = 800$ versus H_a: $\mu > 800$.
$z_{obs} = (848 - 800)/(200/\sqrt{45}) = 1.61$, p-value = .0537
Based on 45 randomly selected property owners there is moderately significant evidence to reject the government's claim.

8.109 H_0: $\mu = 36$ versus H_a: $\mu < 36$.
$t_{obs} = (33.6 - 36)/(4.8/\sqrt{36}) = -3$, p-value = .0025
Based on a random sample of 36 women there is highly significant evidence that women on the pill have a smaller maximal oxygen uptake than women not on the pill.

8.111 H_0: $\mu = 65.42$ versus H_a: $\mu < 65.42$.
$t_{obs} = (64.82 - 65.42)/(2.32/\sqrt{144}) = -3.1$, p-value = .0012
Based on a random sample of 144 adults there is highly significant evidence that the mean adult height of residents in the depressed area is below that of all residents.

8.113 H_0: $\theta = .5$ versus H_a: $\theta > .5$
Sign test of median = 0.5000 versus G.T. 0.5000

	N	BELOW	EQUAL	ABOVE	P-VALUE	MEDIAN
weight	6	2	0	4	0.3437	0.5300

There is insufficient evidence that the produced stones have a median of more than .5 carats.

8.115 H_0: $\pi \geq .7$ versus H_a: $\pi < .7$
$z_{obs} = (.66 - .7)/(\sqrt{(.7)(.3)/200}) = -1.23$, p-value = .1093
Based on a random sample of 200 people there is insufficient evidence to conclude that less than 70% favor raising the drinking age.

8.117 H_0: $\theta \leq 98.5$ versus H_a: $\theta > 98.5$
Sign test of median = 98.50 versus G.T. 98.50

	N	BELOW	EQUAL	ABOVE	P-VALUE	MEDIAN
time	9	1	0	8	0.0195	101.4

T = 8, Based on 9 trials there is significant evidence that the median time is more than 98.5 seconds, therefore you reject the horse. Because his lifetime median time is 100 seconds you did not make an error.

8.119 The stem and leaf plot, boxplot and normal probability plot support the assumptions for a t-test.
H_0: $\mu = 70$ versus H_a: $\mu < 70$.
$t_{obs} = (69.5 - 70)/(4.58/\sqrt{18}) = -.46$, p-value = .3257

Based on the results obtained from 18 volunteers there is insignificant evidence that the counseling process reduces one's score on the exam.

8.121 a   The histogram is skewed right.

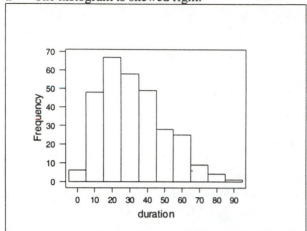

b   The boxplot and normal probability plot show that the distribution is skewed right.

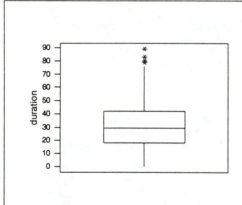

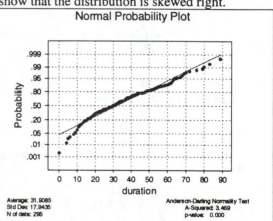

c   Based on the skewness of the distribution the population median is a better representation of a typical incubation period.

d   **T-Test of the Mean**

Test of mu = 30.00 vs mu > 30.00

Variable	N	Mean	StDev	SE Mean	T	P-Value
duration	295	31.91	17.94	1.04	1.83	0.034

Based on these results there is significant evidence that the mean incubation period exceeds 30 months.

e   **Sign test of median = 24.00 versus G.T. 24.00**

	N	BELOW	EQUAL	ABOVE	P-VALUE	MEDIAN
duration	295	114	7	174	0.0003	29.00

There is highly significant evidence that the median incubation period exceeds 2 years.

f   Because of the skewness in the distribution the inference based on the median is more appropriate. Indications are that the median incubation period exceeds 2 years.

8.123 a   The histogram is highly skewed left.

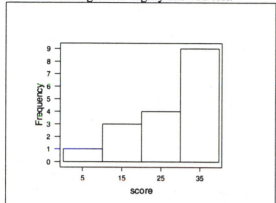

b   The boxplot and normal probability plot also show a skewed left distribution.

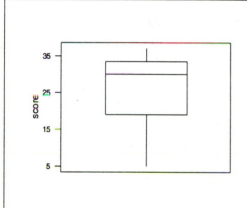 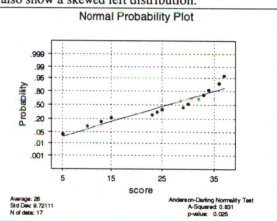

c   The median exam best represents a typical measurement.

d H_0: $\theta \leq 22$ versus H_a: $\theta > 22$

e   **Sign Test for Median**

Sign test of median = 22.00 versus G.T. 22.00

score	N	BELOW	EQUAL	ABOVE	P-VALUE	MEDIAN
	17	4	0	13	0.0245	30.00

Based on the test scores of these 17 patients there is evidence that the tranquilizer significantly improved the amount of learning exhibited by schizophrenics.

8.125 a   The histogram indicates that the distribution may be bimodal.

273

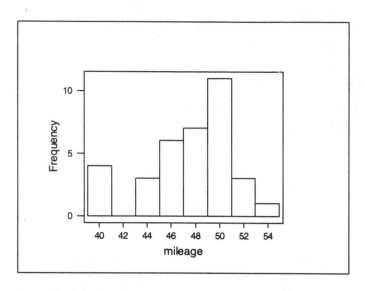

b   The boxplot does not show any unusual behavior or outliers and the normal probability plot
     suggests that the data are reasonably symmetric.

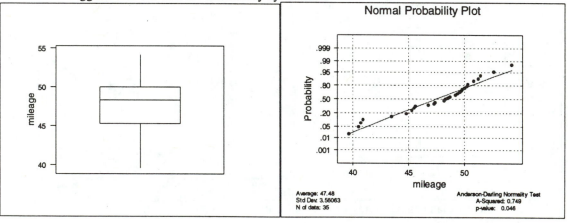

c H_0: $\mu \geq 49$ versus H_a: $\mu < 49$.
d   Because the data are close to symmetric without outliers the test statistic is $\bar{y}$.
e   **T-Test of the Mean**

     Test of mu = 49.000 vs mu < 49.000

Variable	N	Mean	StDev	SE Mean	T	P-Value
mileage	35	47.480	3.561	0.602	-2.53	0.0082

Based on these results there is significant evidence to reject the EPA claim. There is evidence
that the mean miles per gallon is less than 49 in city driving.

9

Inference About The Difference Between Two Parameters

Section 9.1 Introduction

9.1   a   Parameters under study are the population medians.
$H_0: \theta_1 = \theta_2$ versus $H_a: \theta_1 \neq \theta_2$
   b   Parameters under study are the population means.
$H_0: \mu_1 = \mu_2$ versus $H_a: \mu_1 < \mu_2$
μ_1 = mean biodegradation rate for plastic and μ_2 = mean biodegradation rate for paper.
   c   Parameters under study are the population proportions.
$H_0: \pi_1 = \pi_2$ versus $H_a: \pi_1 < \pi_2$.
π_1 = percent covered last year and π_2 = percent covered this year

9.3   Controlled experiment

9.5   The null hypothesis should be rejected. Based on 100 men and women assigned to an experimental and control group there is statistical evidence that mean resting pulse of those who jog over a period of six months is different from those who do not jog.

9.7   To compare the relative danger of the two sports we need the number of participants as well as the number of injuries. Parameters under study are the population proportions.
$H_0: \pi_1 = \pi_2$ versus $H_a: \pi_1 > \pi_2$.
π_1 = proportion of injuries from in-line skating and π_2 = proportion of injuries from skate-boarding

9.9   Subjects were not randomly assigned to the female electrical workers group; this is an observational study. The study does not prove that high doses of electricity causes breast cancer.

9.11 $H_0: \mu_1 = \mu_2$ versus $H_a: \mu_1 \neq \mu_2$
μ_1 = the mean qualifying time for inexperienced drivers and μ_2 = the mean qualifying time for experienced drivers.

Character Dotplot

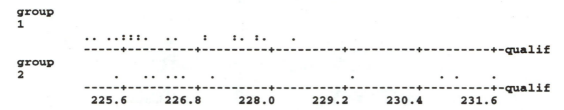

```
group
1
       ..  ..:::.   ..    :    :. :.     .
   ----+---------+---------+---------+---------+---------+-qualif
group
2
           .    .. ...      .            .        .  .    .
   ----+---------+---------+---------+---------+---------+-qualif
      225.6     226.8     228.0     229.2     230.4     231.6
```

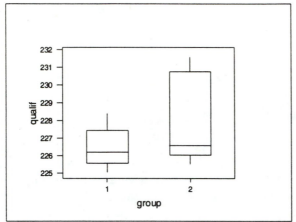

Group 1 is reasonably symmetric and possibly normally distributed. Group 2 is much more variable and possibly bimodal. A group of experienced drivers did no better than the inexperienced drivers but then a group of experienced drivers did significantly better than the inexperienced drivers.

Section 9.2 Inference About the Difference Between Two Population Proportions

9.13   a $p_1 = 65/200 = .325$, $p_2 = 74/200 = .370$, and $p = 139/400 = .3475$

     b $\sqrt{[(.325)(.675)/200 + (.37)(.63)/200]} = .0476$

     c $\sqrt{(.3475)(.6525)}\sqrt{(1/200 + 1/200)} = .0476$

     d   part c

     e

$$z_{obs} = \frac{.325 - .370}{\sqrt{(.3475)(.6525)}\sqrt{(1/200 + 1/200)}} = -.95$$

p-value = 2(.1711) = .3422.

     f   The fact that the data represent the percentages of men and women students with registered cars has no effect on the analysis of data. It affects the conclusion, however, because the conclusion should be written in the context of the problem. In other words, based on the provided data there is insufficient evidence that the percentages of men and women students with registered cars are different.

9.15 a $p_A = 19/250 = .076$ and $p_B = 27/300 = .09$

b   For a confidence interval the sample proportions should not be pooled to estimate the standard error.

c $(.076 - .090) \pm 2.58\sqrt{[(.076)(.924)/250 + (.090)(.910)/300]}$
The confidence interval is -.075 to .047.

d   The interval contains both positive and negative values which means that a two-tailed test of hypothesis would not detect a significant difference between the two proportions.

9.17 p_1 estimates π_1 and p_2 estimates π_2. In hypothesis testing we evaluate the standard error of $p_1 - p_2$ under the assumption that the null hypothesis is true. Because $\pi_1 = \pi_2$ we have that p_1 and p_2 are estimating the same thing; therefore, we pool the two estimates together to get a better estimate of the common value of π. In confidence interval estimation we do not assume that $\pi_1 = \pi_2$ and therefore we do not pool the estimates together.

9.19 H_0: $\pi_1 = \pi_2$ versus H_a: $\pi_1 < \pi_2$.

$$z_{obs} = \frac{.125 - .2}{\sqrt{(.17)(.83)}\sqrt{1/120 + 1/180}} = -1.69; \text{ p-value} = .0455$$

There is significant evidence that the vaccine was effective in reducing the mortality rate. This is a controlled experiment.

9.21 This should be a one-tailed test
H_0: $\pi_1 = \pi_2$ versus H_a: $\pi_1 > \pi_2$.

$$z_{obs} = \frac{.768 - .719}{\sqrt{(.726)(.274)}\sqrt{1/190 + 1/1060}} = 1.40; \text{ p-value} = .0808$$

There is mildly significant evidence that the course was beneficial in increasing reading scores.

9.23 H_0: $\pi_1 = \pi_2$ versus H_a: $\pi_1 \neq \pi_2$.
$\pi_1 =$ proportion of people living in cities who are worried about being a victim of crime and $\pi_2 =$ proportion of people living in suburbs who are worried about being a victim of crime.

$$z_{obs} = \frac{.59 - .57}{\sqrt{(.58)(.42)}\sqrt{1/250 + 1/250}} = .45; \text{ p-value} = 2(.3264) = .6528$$

There is an insignificant difference between the percents of people living in cities and suburbs who are worried about being a victim of crime.

9.25 H_0: $\pi_1 = \pi_2$ versus H_a: $\pi_1 > \pi_2$.
$\pi_1 =$ proportion of female doctors under the age of 35 and $\pi_2 =$ proportion of female doctors between the ages of 35 and 44

$$z_{obs} = \frac{.302 - .224}{\sqrt{(.255)(.745)}\sqrt{1/133718 + 1/198257}} = 50.57; \text{ p-value} < .00001$$

Based on the provided data there is highly significant evidence that more women are entering the medical profession now than in the past.

9.27 H_0: $\pi_1 = \pi_2$ versus H_a: $\pi_1 \neq \pi_2$.
π_1 = proportion of eleventh grade girls who say they are good in math and π_2 = proportion of eleventh grade boys who say they are good in math.

$$z_{obs} = \frac{.48 - .60}{\sqrt{(.549)(.451)}\sqrt{(1/150 + 1/200)}} = -2.23; \text{ p-value} = 2(.0129) = .0258$$

Based on random samples of 150 eleventh grade girls and 200 eleventh grade boys there is highly significant evidence that the percents of eleventh grade girls and boys who say they are good in math are different.

Section 9.3 Inference About the Difference Between Two Population Means

9.29 a $37.4 \pm 1.671*\sqrt{(2366.8/65 + 23357/62)}$ The confidence interval is 3.44 to 71.36
      b $-1.5 \pm 2.457*\sqrt{(6.5/33 + 55.8/34)}$ The confidence interval is -4.83 to 1.83
      c   No, interval b is more valid; it is a 98% confidence interval where as, interval a is a 90% confidence interval.

9.31 a
$$t_{obs} = \frac{6.15 - 8.65}{\sqrt{17.23^2/53 + 8.16^2/48}} = -.94$$
      b   This is a one-tailed test. df = 47, p-value >.15
      c   There is insufficient evidence to reject H_0 and conclude that $\mu_1 - \mu_2 < 0$.

9.33 H_0: $\mu_1 = \mu_2$ versus H_a: $\mu_1 > \mu_2$
μ_1 = the mean annual income in District I and μ_2 = the mean annual income in District II.
$$t_{obs} = \frac{28.65 - 25.94}{\sqrt{9.38^2/38 + 5.47^2/42}} = 1.56, .05 < \text{p-value} < .10$$
There is mildly significant evidence to conclude that the mean annual income in District I is greater than the mean annual income in District II.

9.35 a   Neither sample indicates that inferences should not be based on $\bar{y}_1 - \bar{y}_2$.
      b H_0: $\mu_1 = \mu_2$ versus H_a: $\mu_1 \neq \mu_2$
           μ_1 = the mean grade for a 9am class and μ_2 = the mean grade for a 2pm class
      c
$$t_{obs} = \frac{81.0 - 82.5}{\sqrt{9.56^2/36 + 7.29^2/32}} = -.73$$
      d   p-value > .40 There is insufficient evidence to conclude that the mean grade for a 9am class is different from the mean grade for a 2pm class.

9.37 a H_0: $\mu_1 = \mu_2$ versus H_a: $\mu_1 \neq \mu_2$
     μ_1 = the mean number of errors with drug and μ_2 = the mean number of errors with placebo

   b The parent distributions should not deviate substantially from normality and the samples should be independent.

   c $$t_{obs} = \frac{18.67 - 10.3}{\sqrt{11.21^2/12 + 4.13^2/16}} = 2.46$$

   d df = 11, .02 < p-value < .04 There is significant evidence to conclude that the mean number of errors with the drug is different from the mean number of errors with a placebo.

9.39 a H_0: $\mu_1 = \mu_2$ versus H_a: $\mu_1 > \mu_2$
     μ_1 = mean score of the over 35 age group and μ_2 = mean score of the under 35 age group

   b The boxplots show no significant departures from the basic assumptions for inferences based on $\bar{y}_1 - \bar{y}_2$

   c $$t_{obs} = \frac{27.59 - 24.13}{\sqrt{14.04^2/34 + 10.32^2/32}} = 1.15$$

   Using the conservative approach we get df = 31 and .10 < p-value < .15. Using the more exact method of calculating degrees of freedom we get df = 60 and the same p-value. In either case   there is insignificant evidence to conclude that the mean for the over-35 group is greater than the mean for the under-35 group.

9.41 Both distributions are close to symmetrical. The variability of the two groups is about the same leading to the conclusion that the two groups are homogeneous. The middle 50% of group B is completely contained within the middle 50% of group A. It is doubtful that there is a significant difference between the two groups.

9.43 a Because of the possible bimodality of group 2 there are some doubts about whether the two-sample t-test is appropriate for these data. With such small samples, however, it is difficult to discern a shape for the population distributions. With caution one may conduct a t-test.

   b **Two Sample T-Test and Confidence Interval**

   Two sample T for qualif

group	N	Mean	StDev	SE Mean
2	11	227.86	2.32	0.70
1	22	226.45	1.04	0.22

   95% CI for mu (2) - mu (1): (-0.19, 3.01)
   T-Test mu (2) = mu (1) (vs not =): T= 1.92 P=0.079 DF= 12
   Based on the t-test there is mildly significant evidence that the more experienced drivers have better qualifying times.

9.45 a Both distributions appear skewed right to the point that the two-sample t-procedures may not apply.

   b The mean is much larger than the median because of the right skewness of the distribution.

   c Because of the similar skewness a comparison of population medians is more appropriate for these data.

279

d   The boxplots show a distinct difference between the two distributions suggesting that the median survival time for those in group 1 exceeds the median survival time for those in group 2.

Section 9.4 Inference About the Difference Between Two Population Centers when the Variances are Equal

9.47   a $H_0: \mu_1 = \mu_2$ versus $H_a: \mu_1 < \mu_2$
μ_1 = the mean amount of coffee dispensed by vending machine A and μ_2 = the mean amount of coffee dispensed by vending machine B
b   For small samples the parent populations should be close to a normal distribution and the population variances should be homogeneous.
c $$t_{obs} = \frac{9.8 - 10.1}{1.11\sqrt{(1/10 + 1/12)}} = -0.63 \quad \text{p-value} > .20$$
Insufficient evidence exists to support the conjecture that machine A dispenses significantly less than machine B.

9.49 $(82.4 - 84.2) \pm 1.645(11.5)\sqrt{1/150 + 1/150}$ The confidence interval is (-3.98, .38)
The interval contains both positive and negative values. This means that a two-tailed test of the equality of mean grades for fall and spring classes would not be rejected.

9.51   a   Based on the boxplots there does not seem to be a difference between the mean abilities of men and women.
b   **98% CI for mu female - mu male: (-4.2,   5.1)**
c   The interval contains both positive and negative values. This suggests that there is no significant difference between the abilities of men and women.
d   Because of the long-tailed distributions a confidence interval based on trimmed means might be more appropriate.

9.53   a $H_0: \mu_1 = \mu_2$ versus $H_a: \mu_1 < \mu_2$
μ_1 = the mean carbon monoxide emitted by manufacturer and μ_2 = the mean carbon monoxide emitted by competitor
b   Both distributions appear to be short-tailed.
c   The variances seem to be homogeneous.
d   Neither distribution is severely skewed nor has unusually long tails (they are short-tailed); the pooled t-test can be applied to these data.
e   **Two Sample T-Test and Confidence Interval**
Two sample T for manufac vs compet

	N	Mean	StDev	SE Mean
manufac	9	2.989	0.389	0.13
compet	10	3.370	0.395	0.12

95% CI for mu manufac - mu compet: (-0.76, -0.00)
T-Test mu manufac = mu compet (vs <): T= -2.12   P=0.025   DF=   17
Both use Pooled StDev = 0.392

280

Based on these data there is a significant difference between the mean carbon monoxide emitted by the manufacturer and by the competitor.

9.55   a $H_0: \mu_1 = \mu_2$ versus $H_a: \mu_1 \neq \mu_2$
μ_1 = the mean number of defective gears by A and μ_2 = the mean mean number of defective gears by B

b   Both the normality and the homogeneous variances assumptions seem to be met.

c   **Two Sample T-Test and Confidence Interval**

```
Two sample T for A vs B
       N      Mean     StDev    SE Mean
A  20       23.80      7.32       1.6
B  20       28.75      6.59       1.5

95% CI for mu A - mu B: ( -9.4,   -0.5)
T-Test mu A = mu B (vs not =): T= -2.25   P=0.031   DF=  38
Both use Pooled StDev = 6.97
```
Based on a p-value = .031 there is a significant difference between the number of defective gears produced by the two manufacturers.

9.57   a   Inferences should be on the population medians.

b $H_0: \theta_1 = \theta_2$ versus $H_a: \theta_1 > \theta_2$
θ_1 = median survival time under Arm A and θ_2 = median survival time under Arm B

c   The pooled t-test applied to the ranks is equivalent to the Wilcoxon rank sum test. We have

```
Two Sample T-Test and Confidence Interval

Twosample T for ranks
group   N       Mean     StDev    SE Mean
1       62      71.2     32.4       4.1
2       59      50.3     34.8       4.5

95% C.I. for mu 1 - mu 2: ( 8.7,   33.0)
T-Test mu 1 = mu 2 (vs >): T= 3.41   P=0.0004   DF=  119
Both use Pooled StDev = 33.6
```
Based on the provided data there is highly significant evidence that the median survival time under Arm A is greater than the median survival time under Arm B.

9.59 $H_0: \mu_1 = \mu_2$ versus $H_a: \mu_1 \neq \mu_2$
μ_1 = the mean LDL cholesterol for the placebo group and μ_2 = the mean LDL cholesterol for the treatment group. Conditions are satisfied for the pooled t-test. We have

```
Two Sample T-Test and Confidence Interval

Twosample T for placebo vs treatmen
              N      Mean     StDev    SE Mean
placebo    20      67.2     14.0       3.1
treatmen    9      52.2     17.1       5.7

95% C.I. for mu placebo - mu treatmen: ( 2.7,   27.3)
T-Test mu placebo = mu treatmen (vs not =): T= 2.49   P=0.019   DF=  27
Both use Pooled StDev = 15.0
```

There is significant evidence that the drug reduced the LDL cholesterol level in the treatment group.

Without removing the outlier the results are as follows:

```
Two Sample T-Test and Confidence Interval

Twosample T for placebo vs treatmen
                N       Mean      StDev    SE Mean
placebo        20       67.2      14.0        3.1
treatmen       10       62.2      35.4         11

95% C.I. for mu placebo - mu treatmen: ( -13.4,  23)
T-Test mu placebo = mu treatmen (vs not =): T= 0.56  P=0.58  DF=  28
Both use Pooled StDev = 23.1
```
The one outlier causes the results to be insignificant.

Section 9.5 Inference About the Difference Between Two Population Centers Using Matched Samples

9.61 a The samples are matched because the same subject is used before and after the two-week course of instruction.

b The normal probability plot shows no significant departures from normality.

c $H_0: \mu_d \leq 0$ versus $H_a: \mu_d > 0$ where $\mu_d = \mu_1 - \mu_2$
μ_1 = mean spelling score after course of instruction, μ_2 = mean spelling score before course of instruction

$t_{obs} = 3/(3.94/\sqrt{9}) = 2.28$ p-value = .026

There is significant evidence of improvement after taking the course of instruction.

9.63 a The samples are matched because the same subject is measured before and after viewing the movie.

b The distribution of difference scores should be close to normal. There may be some departure from normality but the sample size is so small it is difficult to tell. There appear to be no gross departures that would preclude the paired t-test.

c A 99% confidence interval for μ_d is

```
Confidence Intervals

Variable      N      Mean     StDev   SE Mean        99.0 % C.I.
differ       12     0.667     1.155     0.333   ( -0.369,   1.702)
```
d The confidence interval contains both positive and negative values, thus there is no change in moral attitudes after viewing the film.

9.65   a   The samples are matched because identical twins are used in the study.

    b   **Character Dotplot**

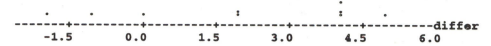

The dotplot shows some departure from normality but with so few observations it is difficult to tell.

    c   **T-Test of the Mean**

Test of mu = 0.000 vs mu not = 0.000

Variable	N	Mean	StDev	SE Mean	T	P-Value
differ	9	2.000	2.500	0.833	2.40	0.043

There is significant evidence that the drug affected intelligence test scores.

9.67   a   There are two extreme outliers on either tail of the distribution that would question the normality assumption.

    b   The normality assumption is doubtful but symmetry is a reasonable assumption, therefore the Wilcoxon signed rank test is the appropriate test.

    c $H_0: \mu_d \le 0$ versus $H_a: \mu_d > 0$ where $\mu_d = \mu_1 - \mu_2$
μ_1 = mean reading comprehension score after the reading course, μ_2 = mean reading comprehension score before the reading course
For the matched pairs t-test the p-value = .17. For the Wilcoxon signed rank test the p-value = .0075. The outliers tend to inflate the standard deviation to the point that the t-test fails to detect a significant difference. The Wilcoxon test is resistant to outliers and detects a significant difference.

    d   Based on the Wilcoxon test there is highly significant evidence that the course was beneficial in raising the reading comprehension scores.

9.69   a   **Character Dotplot**

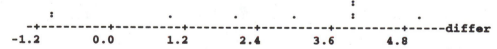

The normality assumption may be in question, the Wilcoxon test may be more appropriate for these data.

    b   **T-Test of the Mean**

Test of mu = 0.00 vs mu not = 0.00

Variable	N	Mean	StDev	SE Mean	T	P-Value
sgnrnks	10	4.70	4.18	1.32	3.56	0.0062

There is highly significant evidence that the mean number of errors is different before and after the course

9.71 a **Stem-and-leaf of cross   N = 15**
         Leaf Unit = 0.10

```
   2    12 00
   2    13
   2    14
   2    15
   2    16
   2    17
   3    18 2
   4    19 1
   5    20 3
  (4)   21 0056
   6    22 011
   3    23 025
```

Stem-and-leaf of self   N = 15
Leaf Unit = 0.10

```
   1    12 7
   1    13
   1    14
   3    15 25
   5    16 25
   6    17 3
  (6)   18 000366
   3    19
   3    20 003
```

The two distributions are similarly skewed left.

b The samples are matched because Darwin planted pairs of seedlings on opposite sides of pots.

c **Stem-and-leaf of differ   N = 15**
   Leaf Unit = 1.0

```
   1    -0 8
   2    -0 6
   2    -0
   2    -0
   2    -0
   5     0 011
  (5)    0 22333
   5     0 5
   4     0 677
   1     0 9
```

There is some doubt about the normality assumption and therefore the data are more suited to the Wilcoxon singed rank test.

d **T-Test of the Mean**

Test of mu = 0.00 vs mu not = 0.00

Variable	N	Mean	StDev	SE Mean	T	P-Value
sgnrnks	15	4.80	7.99	2.06	2.33	0.036

There is a significant difference in the heights of the plants from the two methods of fertilization.

Section 9.6 Robust Inference About the Difference Between Two Population Means

9.73 a
$$z_{obs} = \frac{(55.8 - 62.4)}{\sqrt{(242.1/27 + 236.5/27)}} = -1.57; \text{ p-value} = .0582$$

There is mildly significant evidence that the difference between μ_1 and μ_2 is less than 0.

b
$$z_{obs} = \frac{(467 - 482)}{\sqrt{(6241/96 + 4225/107)}} = -1.47; \text{ p-value} = 2(.0708) = .1416$$

There is insufficient evidence to conclude that the dfference between μ_1 and μ_2 is not 0.

9.75 H_0: $\mu_1 = \mu_2$ versus H_a: $\mu_1 \neq \mu_2$
$$z_{obs} = \frac{(51.556 - 48.389)}{\sqrt{(29.47/18 + 37.28/18)}} = 1.64; \text{ p-value} = 2(.0505) = .101$$

There is insufficient evidence to conclude that the means are different.

9.77 H_0: $\mu_1 \leq \mu_2$ versus H_a: $\mu_1 > \mu_2$
$$z_{obs} = \frac{(87.83 - 76.96)}{\sqrt{(5.97^2/24 + 11.3^2/24)}} = 4.17; \text{ p-value} = .00003$$

There is highly significant evidence that the mean score for the computer group exceeds the mean score for the non-computer group.

9.79 H_0: $\mu_1 = \mu_2$ versus H_a: $\mu_1 \neq \mu_2$
$$z_{obs} = \frac{(72.167 - 72.125)}{\sqrt{(28.2/24 + 31.9/24)}} = 0.03; \text{ p-value} = .976$$

There is insufficient evidence that the mean score for the learning disabled students is different from the mean score for normal achievers.

Section 9.8 Summary and Review

9.81 a   The standard deviations are close enough that we can pool them together to estimate the standard error of $\bar{y}_1 - \bar{y}_2$.

b $13 \pm 1.66*30.71\sqrt{(1/65 + 1/60)}$ The confidence interval is 3.87 to 22.13

c   The interval contains only positive values. This means that a two-tailed test of the equality of the population means would be rejected at the 10% level of significance. The mean size in Region one is greater than the mean size in Region two.

9.83 H_0: $\theta_1 = \theta_2$ versus H_a: $\theta_1 \neq \theta_2$
Twosample T for ranks
```
group    N       Mean       StDev    SE Mean
1        21      23.7       10.6       2.3
2        19      17.0       12.1       2.8
```
95% C.I. for mu 1 - mu 2: (-0.5, 14.0)
T-Test mu 1 = mu 2 (vs not =): T= 1.87 P=0.069 DF= 38
Both use Pooled StDev = 11.3
There is a mildly significant difference between the two population medians.

9.85 a   Matched pairs t-test.
      b   The assumptions are not violated.
      c H_0: $\mu_d \leq 0$ versus H_a: $\mu_d > 0$ where $\mu_d = \mu_1 - \mu_2$
           μ_1 = mean yield for the new variety, μ_2 = mean yield for the standard variety.
      d   **Test of mu = 0.00 vs mu > 0.00**

```
Variable      N       Mean      StDev     SE Mean        T      P-Value
differ        12      7.75      7.01      2.02          3.83     0.0014
```
      e   There is highly significant evidence that the yield per acre for the new variety is higher
           than the standard variety.

9.87 a   The assumptions are not violated.
      b   The samples are independent.
      c $(17.63 - 14.91) \pm 2.467 * 4.36\sqrt{(1/15 + 1/15)}$ The confidence interval is -1.21 to 6.65
      d   The 98% confidence interval contains both positive and negative values. A null hypothesis
           of the equality of the population means cannot be rejected when the level of significance is
           2%. Based on these results we would say that there is an insignificant difference between
           the number of hours spent on homework by students attending public and private high
           schools.

9.89 $134.6 - 127.8 \pm 1.96 * \sqrt{(26.82^2/32 + 23.47^2/34)}$ The interval is   -5.39 to 18.99
      The confidence interval based on trimmed means is recommended when the parent distributions
      are symmetric with long tails.

9.91 $(.966 - .448) \pm 1.96\sqrt{[(.966)(.034)/29 + (.448)(.552)/143]}$ The confidence interval is .413 to .623

9.93 H_0: $\mu_d \leq 0$ versus H_a: $\mu_d > 0$ where $\mu_d = \mu_1 - \mu_2$
      μ_1 = mean number of sit-ups after the course, μ_2 = mean number of sit-ups before the course
      A matched pairs t-test:
      T-Test of the Mean

      Test of mu = 0.000 vs mu > 0.000

```
Variable      N       Mean      StDev     SE Mean        T      P-Value
differ        9       2.000     2.179     0.726         2.75     0.012
```
      There is significant evidence that the physical fitness course improved the number of sit-ups by
      the participants.

9.95 H_0: $\pi_1 = \pi_2$ versus H_a: $\pi_1 > \pi_2$.

π_1 = proportion interested in business and tech now and π_2 = proportion interested in business and tech in the past.

$$z_{obs} = \frac{.6275 - .59}{\sqrt{(.60875)(.39125)}\sqrt{(1/400 + 1/400)}} = 1.09; \text{ p-value} = .1379$$

There is insignificant evidence that a higher proportion of students are interested in business and tech now than in the past.

9.97 H_0: $\pi_1 = \pi_2$ versus H_a: $\pi_1 > \pi_2$.

π_1 = proportion from the hyperactive group and π_2 = proportion from the normal group.

$$z_{obs} = \frac{.26 - .16}{\sqrt{(.22)(.78)}\sqrt{(1/301 + 1/191)}} = 2.61; \text{ p-value} = .0045$$

There is highly significant evidence that a higher proportion of hyperactive children had mothers with poor health during pregnancy.

9.99 **T-Test of the Mean**

Test of mu = 0.00 vs mu not = 0.00

Variable	N	Mean	StDev	SE Mean	T	P-Value
differ	14	3.86	6.29	1.68	2.30	0.039

Based on the p-value there is a significant difference between the test scores before and after the vocabulary training. The following normal probability plot does not rule out normality and therefore the matched pairs t-test is appropriate for analysis of these data.

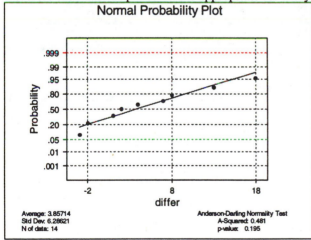

9.101 The following descriptive statistics show that the standard deviations are close enough to assume the variances are homogeneous. The side-by-side boxplots show that the distributions are symmetric without outliers. A pooled t-test can be used to compare the population means.

```
Descriptive Statistics

Variable        N        Mean     Median   TrMean   StDev    SEMean
Fourth          10       112.80   112.00   112.50   6.86     2.17
Colleag         10       112.10   112.50   112.13   9.79     3.10

Variable        Min      Max         Q1       Q3
Fourth          104.00   124.00      106.50   119.50
Colleag         99.00    125.00      102.25   121.50
```

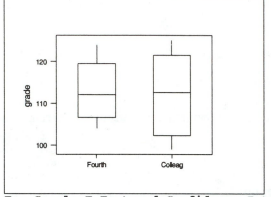

Two Sample T-Test and Confidence Interval

```
Twosample T for Fourth vs Colleag
          N       Mean      StDev    SE Mean
Fourth    10      112.80    6.86     2.2
Colleag   10      112.10    9.79     3.1

95% C.I. for mu Fourth - mu Colleag: ( -7.2,   8.6)
T-Test mu Fourth = mu Colleag (vs >): T= 0.19  P=0.43  DF=  18
Both use Pooled StDev = 8.45
```
Based on the p-value there is an insignificant difference between the spelling grades for the two classes. There is no evidence to support the fourth grade teacher's claim.

9.103 The accompanying boxplot indicates no significant departures from assumptions for the matched pairs t-test.

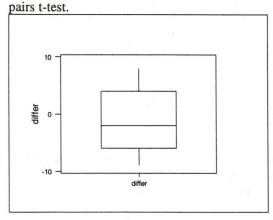

```
T-Test of the Mean

Test of mu = 0.00 vs mu not = 0.00

Variable      N        Mean      StDev     SE Mean         T      P-Value
differ        15      -1.00       5.68        1.47     -0.68         0.51
```
The difference between the readings is insignificant. To check the accuracy of a machine multiple readings would have to be taken on the same subject.

9.105 The following boxplot of the difference scores identifies one outlier but otherwise the distribution does not deviate substantially from the assumptions for the matched pairs t-test.

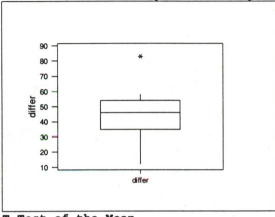

```
T-Test of the Mean

Test of mu =  0.00 vs mu not =  0.00

Variable      N        Mean      StDev     SE Mean         T      P-Value
differ        11      45.18      17.62        5.31      8.51       0.0000
```
There is highly significant evidence that the arterial blood pressure is different after receiving oxytocin.

Unit 3 [Chapters 7-9]

Significant Ideas and Review Exercises

U3.1   a   No, the formulas in Chapter 7 apply to simple random samples. This is a stratified random sample.

      b   The statement attempts to relate sample size and confidence level. However, no mention is made of the margin of error. In confidence interval estimation we are attempting to estimate a certain population parameter to within a specified margin of error. The sample size is related to the confidence level, but it is also indirectly related to the margin of error. To decrease the margin of error the sample size must increase. A correct statement is something like, "Enough responses were obtained to be 95% confident that the estimate is within a specified margin of error of the true population parameter."

U3.3   a $n = (z^*\sigma/B)^2 = (2.58*27/10)^2 \approx 49$

      b $128 \pm 2.58(27)/\sqrt{45}$ The confidence interval is 117.62 to 138.38.

      c   The value \$135 is inside the 99% confidence interval and therefore the null hypothesis $H_0: \mu = 135$ would not be rejected at the 1% level of significance.

U3.5   a $n = (2.33*.5/.03)^2 \approx 1509$

      b $p = 94/500 = .188$

         $.188 \pm 2.33\sqrt{(.188)(.812)/500}$ The confidence interval is .147 to .229.

      c   The hypothesis $H_0: \pi = .15$ would not be rejected with $\alpha = .02$ because .15 is inside the 98% confidence interval.

U3.7 $51 \pm 2.626(16)/\sqrt{100}$ The confidence interval is 46.8 to 55.2

U3.9 $H_0: \mu \leq 450$ versus $H_a: \mu > 450$

      $t_{obs} = (463 - 450)/(42.6/\sqrt{100}) = 3.05$ $.001 < \text{p-value} < .005$

      There is highly significant evidence that the applicants for admission are above the national norm of 450 on SAT math.

U3.11 $H_0: \mu = 8$ versus $H_a: \mu \neq 8$

      $t_{obs} = (7.85 - 8)/(1.15/\sqrt{400}) = -2.61$ $.001 < \text{p-value} < .005$

      There is highly significant evidence that the mean weight is not 8 ounces. Based on their established criteria the supermarket should reject the shipment.

U3.13 $H_0: \mu_1 \leq \mu_2$ versus $H_a: \mu_1 > \mu_2$

      μ_1 = the mean life span in open environment and μ_2 = the mean life span in caged environment.

$$t_{obs} = \frac{7.6 - 5.9}{\sqrt{1.4^2/35 + 3.7^2/62}} = 3.23, \; .001 < \text{p-value} < .005$$

There is highly significant evidence that the mean life span of animals is greater in an open environment than in a caged environment.

U3.15 The pooled t-test assumes that the population variances are homogeneous and that the parent populations are close to normally distributed. The larger the sample sizes the more the distributions can deviate from normality. However, we should not use a t-test if either distribution is severely skewed or is symmetric with long tails. The boxplot can identify outliers and describe the general shape of the distribution. A normal probability plot checks deviations from normality.

U3.17 a $n = (2.33*.5/.04)^2 \approx 849$

   b $p = 500/900 = .556$

   $.556 \pm 2.33\sqrt{(.556)(.444)/900}$ The confidence interval is .517 to .595. We are 98% confident that the percent of smokers who would light up without asking permission is somewhere between 51.7% and 59.5%.

U3.19 H_0: $\mu \leq 20$ versus H_a: $\mu > 20$

   $t_{obs} = (22 - 20)/(3/\sqrt{16}) = 2.67$ $.005 < \text{p-value} < .01$

   There is highly significant evidence that this group of college administrators has an unusually high self-concept. To conduct a t-test with 16 observations the parent distribution should be symmetric without outliers.

U3.21 An estimate of σ is range/4 = 16/4 = 4 months; $n = (z^*\sigma/B)^2 = (1.96*4/.5)^2 \approx 246$

U3.23

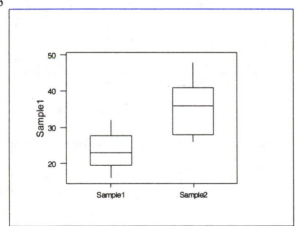

```
Descriptive Statistics
Variable      N      Mean    Median    TrMean    StDev    SEMean
Sample1       8     23.62     23.00     23.62     5.21      1.84
Sample2       7     35.57     36.00     35.57     7.55      2.85
```

Variable	Min	Max	Q1	Q3
Sample1	16.00	32.00	19.50	27.75
Sample2	26.00	48.00	28.00	41.00

The variability exhibited in the boxplots by the two samples suggest that the populations are homogeneous. Neither standard deviation is more than twice the other; a pooled t-test seems appropriate.

```
Twosample T for Sample1 vs Sample2
            N      Mean    StDev   SE Mean
Sample1    8     23.62    5.21      1.8
Sample2    7     35.57    7.55      2.9
```

```
95% C.I. for mu Sample1 - mu Sample2: ( -19.1,  -4.8)
T-Test mu Sample1 = mu Sample2 (vs not =): T= -3.61  P=0.0032  DF=  13
Both use Pooled StDev = 6.39
```

Based on the p-value the difference between the two population means is highly significant. If we examine the boxplots, however, the difference is apparent. The upper quartile of Sample 1 is at the lower quartile of Sample 2. There is a clear distinction between the groups; the formal test of significance was not required.

U3.25 The boxplot and normal probability plot show that the distribution is very close to normal.

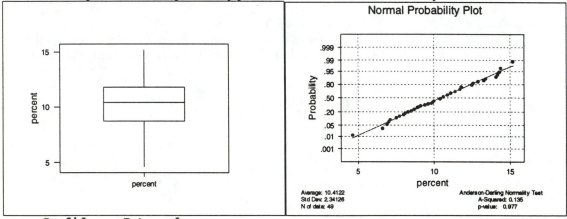

Confidence Intervals

Variable	N	Mean	StDev	SE Mean	99.0 % C.I.
percent	49	10.412	2.341	0.334	(9.515, 11.310)

We are 99% confident that the true percent of high school dropouts is somewhere between 9.5% and 11.3%.

U3.27 The boxplot and normal probability plot indicate that the distribution is symmetric with long tails.

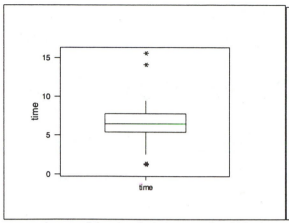
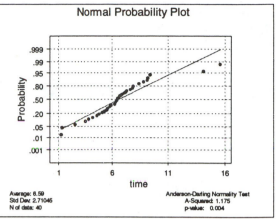

A test of the population mean should be based on $\bar{y}_T$. We have

$H_0: \mu \geq 7$ versus $H_a: \mu < 7$

$n = 40$, $k = 32$; $\bar{y}_{T.10} = 6.463$, $s_T = 1.854$, $z_{obs} = (6.463 - 7)/(1.854/\sqrt{32}) = -1.64$ p-value = .0505

There is mildly significant evidence that the mean checkout time is less than 7 minutes per customer. By comparison the following t-test based on $\bar{y}$ is insignificant.

T-Test of the Mean

Test of mu = 7.000 vs mu < 7.000

Variable	N	Mean	StDev	SE Mean	T	P-Value
time	40	6.590	2.710	0.429	-0.96	0.17

U3.29 The distribution of difference scores appears to be symmetric with very short tails. The matched pairs t-test is appropriate for these data.

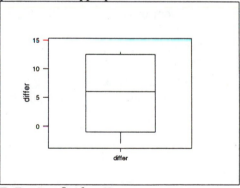

T-Test of the Mean

Test of mu = 0.00 vs mu not = 0.00

Variable	N	Mean	StDev	SE Mean	T	P-Value
differ	9	6.00	6.44	2.15	2.79	0.023

Based on the p-value there is a significant difference between the effectiveness of the two brands of fly spray.

293

10

Analysis of Categorical Data

Section 10.1 The Chi-Square Goodness of Fit

10.1 a   25.0   b 30.6   c   7.26   10.3 4.4 and 23.3

10.5 H_0: $\pi_1 = \pi_2 = \pi_3 = \pi_4 = \pi_5 = 1/5$ versus H_a: at least one is not the same
$\chi^2_{obs} = 7.34$; p-value > .10
There is insufficient evidence to conclude that the classes are not equally likely.

10.7 If H_0: $\pi_1 = .40$, $\pi_2 = .45$, $\pi_3 = .07$, $\pi_4 = .05$, $\pi_5 = .03$ is true, then $e_1 = 480$, $e_2 = 540$, $e_3 = 84$, $e_4 = 60$, $e_5 = 36$ and $\chi^2_{obs} = 5.81$. p-value > .10
There is insufficient evidence to conclude that the percentages are different from those specified by the official.

10.9 H_0: $\pi_1 = \pi_2 = \pi_3 = \pi_4 = \frac{1}{4}$ versus H_a: at least one is not the same
$\chi^2_{obs} = 3.6$; p-value > .10
There is insufficient evidence to conclude that the proportions of marbles are different.

10.11 H_0: The strains are equally resistant to the chemical agent.
H_a: The strains are not equally resistant to the chemical agent.
If H_0 is true then $e_1=e_2=e_3=e_4=e_5 = 200$ and $\chi^2_{obs} = 128.08$; p-value < .005
The evidence is highly significant that some strains are more resistant to the chemical than others.

10.13 If H_0: $\pi_1 = .31$, $\pi_2 = .50$, $\pi_3 = .14$, $\pi_4 = .05$ is true, then $e_1 = 77.5$, $e_2 = 125$, $e_3 = 35$, $e_4 = 12.5$ and
$\chi^2_{obs} = 206.59$ p-value < .005
There is highly significant evidence that the claims from the employees of this company do not follow the same distribution as given in the Wall Street Journal.

10.15 The categories on the survey do not constitute a multinomial experiment and therefore the chi-square goodness of fit does not apply. The categories of a multinomial experiment must be independent and the associated probabilities must sum to unity. This is clearly not the case in this study.

Section 10.2 The Chi-Square Test of Independence

10.17 H_0: Variables A and B are independent.
H_a: Variables A and B are dependent.
$\chi^2_{obs} = 9.597$; df $= 2$, p $= 0.008$
There is highly significant evidence that the two variables are dependent.

10.19 H_0: The service rendered is independent of the type dealership
H_a: The service rendered is dependent on the type dealership
$\chi^2_{obs} = 8.605 + 7.546 + 2.679 + 2.350 + 0.334 + 0.293 + 0.144 + 0.126 + 0.488 + 0.428 + 3.891 + 3.412 = 30.297$
df $= 5$, p $= 0.000$
There is highly significant evidence that the service rendered is dependent on the type dealership.

10.21 H_0: The presence of toxicity and tumor regression are independent.
H_a: The presence of toxicity and tumor regression are dependent.
$\chi^2_{obs} = 5.498 + 3.869 + 4.229 + 2.976 = 16.571$; df $= 1$, p $= 0.000$
There is highly significant evidence that the presence of toxicity and tumor regression are dependent. All expected cell counts are greater than 5. This is important because the χ^2 statistic has an approximate chi-squrare distribution when no more than 20% of the cells have an expected count less than 5. If this condition is not met then the approximation is not good.

10.23 a H_0: The adverse events are independent of the medication.
H_a: The adverse events are dependent on the medication.
$\chi^2_{obs} = 1.252 + 2.579 + 15.358 + 0.895 + 0.488 + 5.873 + 0.185 + 2.746 + 8.456 = 37.832$
df $= 4$, p $= 0.000$
Based on the p-value there is highly significant evidence that the adverse events are dependent on the medication.

b ChiSq $= 0.007 + 0.363 + 0.247 + 0.016 + 0.905 + 0.616 = 2.154$
df $= 2$, p $= 0.341$
Based on this p-value the adverse events and medication are independent.

c The effects of the medications on insomnia and the fact that the placebo had no effect on insomnia are the causes of the dependence found in part a.

d Because headache and drowsiness are independent of the medication the reported claim is statistically valid.

10.25 H_0: A students desire to participate in the science project program is independent of academic standing.
H_a: A students desire to participate in the science project program is dependent on academic standing.
$\chi^2_{obs} = 0.003 + 0.006 + 0.845 + 1.658 + 0.946 + 1.856 = 5.314$; df $= 2$, p $= 0.071$
There is mildly significant evidence that participating in the science project depends on academic standing.

10.27 a H_0: Opinion on embargo against Cuba and date of survey are independent.
          H_a: Opinion on embargo against Cuba and date of survey are dependent.
          $\chi^2_{obs} = 0.239 + 0.234 + 0.651 + 0.637 = 1.762$; df = 1, p = 0.185
      b   The provided data show that the opinion on embargo against Cuba and the date of the survey
          are independent. The numerical gain from December, 1993 to June, 1994 in those favoring
          an embargo against Cuba is statistically insignificant.

10.29 H_0: Injury level and wearing a seatbelt are independent.
      H_a: Injury level and wearing a seatbelt are dependent.
      $\chi^2_{obs} = 4.162 + 12.245 + 30.237 + 3.120 + 0.791 + 2.328 + 5.749 + 0.593 = 59.224$
      df = 3, p = 0.000
      There is highly significant evidence that the injury level is statistically related to whether or not
      the driver wears a seatbelt.

Section 10.3 The Chi-Square Test of Homogeneity

10.31 H_0: The proportions falling in these three categories are the same for all three populations.
      H_a: The proportions falling in these three categories are the different for at least one of these
          populations.
      $\chi^2_{obs} = 4.621$; p-value = .329
      The row marginal totals are all 80; they are fixed values. All of the expected cell counts exceed
      5 (the smallest is 21.33) so the chi-square approximation should be good. Based on the p-value
      there is insufficient evidence that the proportions falling in the 3 categories are different.

10.33 a   The classification variable is the opinion on the referendum.
      b   The populations are the voters from District 1 and from District 2.
      c H_0: The proportion of people having the same opinion is the same for both districts.
          H_a: The proportion of people having the same opinion is different for the two districts.
      d $\chi^2_{obs} = 0.545 + 1.536 + 0.038 + 0.545 + 1.536 + 0.038 = 4.241$; df = 2, p = 0.121
          There is insufficient evidence that the proportions in the two districts are different.
      e   No assumptions for the test have been violated.

10.35 a   The classification variable is the opinion of the respondent.
      b   The populations are the players and members of the press.
      c H_0: Football players and members of the press have the same opinion.
          H_a: Football players and members of the press have different opinions.
          $\chi^2_{obs} = 9.152 + 5.420 + 0.180 + 6.537 + 3.872 + 0.129 = 25.291$; df = 2, p = 0.000
          There is highly significant evidence that football players and the press have different
          opinions on who should receive credit for good play.

10.37 The classification variable is whether or not the subject has contemplated suicide. The populations are Vietnam veterans and non-veterans.

H_0: The proportion who have contemplated suicide is the same for both veterans and non-veterans.

H_a: The proportion who have contemplated suicide is not the same for veterans and non-veterans.

$\chi^2_{obs} = 5.128 + 5.128 + 1.404 + 1.404 = 13.065$; df = 1, p = 0.000

There is highly significant evidence that the proportion of veterans who have contemplated suicide is different from non-veterans.

10.39 a The classification variable is the preference of a favorite sport.

b The populations are collegiate men and women.

c H_0: Favorite sport is distributed the same for men and women.

H_a: Favorite sport is not distributed the same for men and women.

d $\chi^2_{obs} = 0.176 + 2.449 + 1.038 + 7.113 + 0.176 + 2.449 + 1.038 + 7.113 = 21.553$

df = 3, p = 0.000

There is highly significant evidence that favorite sports are not distributed the same for men and women.

10.41 a The classification variable is the injury frequency.

b The populations are 1988-90 and 1991-93 Chevrolet vehicles. It is assumed that samples of vehicles from the two model years are fixed and have been classified into the injury categories.

c

Chi-Square Test

Expected counts are printed below observed counts

	A	B	C	D	F	Total
1	16	5	5	3	4	33
	13.79	3.45	8.37	2.46	4.93	
2	12	2	12	2	6	34
	14.21	3.55	8.63	2.54	5.07	
Total	28	7	17	5	10	67

ChiSq = 0.354 + 0.699 + 1.359 + 0.117 + 0.174 +
         0.343 + 0.678 + 1.319 + 0.114 + 0.169 = 5.326
df = 4, p = 0.256
5 cells with expected counts less than 5.0

From the computer printout we see that 5 of the 10 cells have expected counts less than 5. This may cause serious problems with the chi-square approximation.

d We adjust the data by combining the 1st and 2nd columns and combining the 4th and 5th columns. We now have three classifications: Above average, Average, Below average.

```
Chi-Square Test

Expected counts are printed below observed counts

                A         B         C      Total
        1       21        5         7        33
             17.24     8.37      7.39

        2       14       12         8        34
             17.76     8.63      7.61

Total           35       17        15        67

ChiSq =  0.821 +  1.359 +  0.020 + 0.796 +  1.319 +  0.020 = 4.335
df = 2,  p = 0.115
```

e H_0: Injury frequency is the same for the two model years.
    H_a: Injury frequency is different for the two model years.

f $\chi^2_{obs} = 4.335$; p-value = .115 Based on these data the injury frequency distribution is the same for the two model years.

Section 10.4 Summary and Review

10.43 H_0: The row and column classifications are independent.
    H_a: The row and column classifications are dependent.
    $\chi^2_{obs} = 1.899$; p-value = .594 There is insufficient evidence to conclude that the row and column classifications are dependent. This is the chi-square test of independence.

10.45 H_0: The proportions of suicides are the same for all 4 classes.
    H_a: The proportions of suicides are different for the 4 classes.
    $\chi^2_{obs} = 3.6842$; df = 3, p-value > .10
    There is insufficient evidence to say that the proportion in the 4 classes are different.

10.47 H_0: Moral values and opinion on the referendum are independent.
    H_a: Moral values and opinion on the referendum are dependent.
    $\chi^2_{obs} = 1.472 + 1.118 + 0.037 + 0.325 + 0.314 + 0.002 + 8.998 + 7.363 + 0.074 = 19.702$
    df = 4, p = 0.001
    Their moral values are related to their opinion on the referendum.

10.49 H_0: The geographical distribution of prisoners on death row is the same as the distribution of the general population.
    H_a: The geographical distribution of prisoners on death row is different from the distribution of the general population.
    $\chi^2_{obs} = 644.68$; df = 3, p-value < .001

```
```

There is highly significant evidence that the geographical distribution of prisoners on death row is different from the distribution of the general population.

10.51 H_0: The type of facility performing abortions is the same for 1988 and 1992.
H_a: The type of facility performing abortions is different for 1988 and 1992.
$\chi^2_{obs} = 10.514$; df = 3, p-value = .015
There is significant evidence that the type of facility performing abortions is different for 1988 and 1992.

10.53 H_0: The holdings are equally divided among the three types of stock.
H_a: The holdings are not equally divided among the three types of stock.
$\chi^2_{obs} = 0.93846$; p-value > .1
There is insufficient evidence to conclude that the stocks are not equally divided.

10.55 H_0: The sales are equally distributed among the salespeople.
H_a: The sales are not equally distributed among the salespeople.
$\chi^2_{obs} = 14.435$; .01 < p-value < .025
There is significant evidence that the sales are not equally distributed among the salespeople.

10.57 H_0: Size and outlook are independent.
H_a: Size and outlook are dependent.
$\chi^2_{obs} = 0.097 + 0.003 + 0.275 + 0.200 + 0.756 + 0.116 + 0.011 + 1.205 + 1.246 = 3.910$
df = 4, p = 0.419
There is insufficient evidence that size of corporation and outlook on the economy are related.

10.59 H_0: Sex of dead deer and food supply are independent.
H_a: Sex of dead deer and food supply are dependent.
$\chi^2_{obs} = 0.064 + 0.077 + 0.000 + 0.000 + 0.171 + 0.205 = 0.516$
df = 2, p = 0.772
There is insufficient evidence that an association exists between sex of the deer and the food supply.

10.61 a H_0: The education levels of black women is the same as black men.
H_a: The education levels of black women is different from black men.
b This is a chi-square test of homogeneity.

c **Chi-Square Test**

Expected counts are printed below observed counts

	female	male	Total
1	486	496	982
	553.02	428.98	
2	659	530	1189
	669.59	519.41	
3	691	435	1126
	634.12	491.88	
4	208	134	342
	192.60	149.40	
5	96	65	161
	90.67	70.33	
Total	2140	1660	3800

ChiSq = 8.122 + 10.471 + 0.168 + 0.216 + 5.103 + 6.578 + 1.231 + 1.587 + 0.314 + 0.404 = 34.195

d df = 4, p = 0.000   The null hypothesis should be rejected.

e The evidence is highly significant that the education levels of black women is different from black men.

10.63 H_0: Perceived math ability is the same for girls and boys
H_a: Perceived math ability is different for girls and boys

Chi-Square Test

Expected counts are printed below observed counts

	hopeless	belowavg	average	aboveavg	superior	Total
1	56	61	54	21	8	200
	45.50	52.00	57.50	31.50	13.50	
2	35	43	61	42	19	200
	45.50	52.00	57.50	31.50	13.50	
Total	91	104	115	63	27	400

ChiSq = 2.423 + 1.558 + 0.213 + 3.500 + 2.241 +
        2.423 + 1.558 + 0.213 + 3.500 + 2.241 = 19.869
df = 4, p = 0.001

Based on these data there is highly significant evidence that perceived math ability is different for girls and boys.

10.65 a   The percents converted to counts:

		Wear seat belt
Region	yes	no
East	92	290
Midwest	103	292
South	75	341
West	113	210

Chi-Square Test

Expected counts are printed below observed counts

	C1	C2	Total
1	92	290	382
	96.51	285.49	
2	103	292	395
	99.79	295.21	
3	75	341	416
	105.10	310.90	
4	113	210	323
	81.60	241.40	
Total	383	1133	1516

ChiSq = 0.211 + 0.071 + 0.103 + 0.035 + 8.619 + 2.914 + 12.081 + 4.084 = 28.117
df = 3, p = 0.000

Based on the survey data there is highly significant evidence that the frequency of wearing seatbelts is different for different regions of the country.

b   The percents converted to counts:

	Wear seat belt		
Age		yes	no
under 30	92	237	
30-49	148	421	
50 and older		134	475

Chi-Square Test

Expected counts are printed below observed counts

	C1	C2	Total
1	92	237	329
	81.65	247.35	
2	148	421	569
	141.21	427.79	
3	134	475	609
	151.14	457.86	
Total	374	1133	1507

ChiSq = 1.312 + 0.433 + 0.326 + 0.108 + 1.943 + 0.642 = 4.764
df = 2, p = 0.093

Based on the survey data there is mildly significant evidence that the frequency of wearing seatbelts is different for different age groups.

c   The percents converted to counts:

```
                Wear seat belt
Income          yes      no
level 1         79       147
level 2         55       127
level 3         72       227
level 4         98       329
level 5         52       254
Chi-Square Test
```

```
Expected counts are printed below observed counts

           C1       C2      Total
   1        79      147       226
         55.87   170.13

   2        55      127       182
         44.99   137.01

   3        72      227       299
         73.92   225.08

   4        98      329       427
        105.56   321.44

   5        52      254       306
         75.65   230.35
Total      356     1084      1440
```

```
ChiSq =  9.574 +  3.144 + 2.225 +  0.731 + 0.050 +  0.016 + 0.542 +
0.178 + 7.394 +  2.428 = 26.281
df = 4, p = 0.000
```

Based on the survey data there is highly significant evidence that the frequency of wearing seatbelts is different for different income levels.

10.67   a H_0: The type of drug offense is independent of race.

   H_a: The type of drug offense is dependent on race.

   b   **Chi-Square Test**

```
Expected counts are printed below observed counts

          heroin     crack   cocaine  marijuan     Total
   1         407       106      4525      2825      7863
          806.57    831.80   4558.79   1665.84

   2        1156      2513      4439       442      8550
          877.04    904.48   4957.09   1811.39

   3        1314       348      7297      2675     11634
         1193.39   1230.72   6745.12   2464.76

Total      2877      2967     16261      5942     28047
```

302

```
ChiSq =197.944 +633.309 +  0.250 +806.582 +
       88.729 +2.9E+03 + 54.149 +1.0E+03 +
         12.189 +633.123 + 45.154 + 17.932 = 6385.210
df = 6, p = 0.000
```
The chi-square statistic is extremely large.

c   The null hypothesis should be rejected.

d   The largest differences between actual and expected counts occur in cells with rows intersecting with crack and marijuana.

e   Many more blacks are convicted for a crack drug offense than either whites or hispanics. Very few blacks are convicted for a marijuana drug offense. But a larger proportion of whites are convicted for a marijuana offense.

11

Regression Analysis

Section 11.1 The Linear Regression Model

11.1   a $e = 3 - 17 + 16 = 2$
      b $e = 20 - 17 - 4 = -1$
      c $e = 5 - 17 + 12 = 0$

11.3   a $\mu_y = 65{,}280 + 24.3(1500) = 101{,}730$
      b $101730 \pm 2(2300)$ The interval is (97130, 106330). Out of 100 randomly selected homes we expect 5 to have prices outside this interval.
      c $24.3(100) = 2430$
      d   The cost 97,500 is consistent with the interval found in part b.

11.5   a   The mean cost 3 months after being introduced is $\mu_y = 4800 - 375(3) = 3675$. The initial cost is 4800.
      b $3675 \pm 2(45)$ The interval is (3585, 3765). We expect 5 to have prices outside this interval.
      c   375
      d   After 3 months $3,750 is consistent with the interval found in part b.
      e   The model is proposed for the first six months. There is no reason to believe that the same linear model will apply for a whole year.

11.7

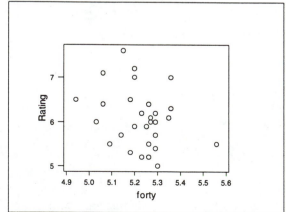

Correlation of Rating and forty = -0.268

The regression equation is Rating = 13.7 - 1.47 forty
SSE = 11.2290, s = 0.6449
The linear relationship between rating and forty is weak.

11.9 This is an association problem. The experimenter is hoping to show an association between two variables. In a comparison problem the goal is to compare two or more groups of subjects. For example, a comparison problem might compare the risk of heart disease for men and women who have high waist-to-hip ratios. On the other hand, an association problem, like this one, is attempting to show a relationship between the risk of heart disease and the waist-to-hip ratio.

11.11 a A straight line could fit the data reasonably well, but a curvilinear fit seems more appropriate. There are no unusual observations other than the curvilinear pattern.
  b The regression equation is: magnesiu = 5.77 - .000252 distance. The estimated standard deviation is s = 1.195. The following regression plot shows that the straight line fits the data reasonably well.

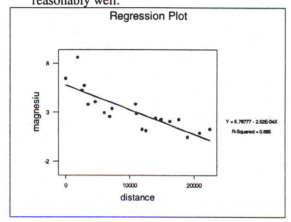

Section 11.2 Inference About the Regression Coefficients

11.13 a s = 1.464
  b $\hat{y} = .783 + .6848 x$
  c $t_{obs} = b_1 / SE(b_1) = .6848/.1526 = 4.49$
  d p-value = .004 There is highly significant evidence that $\beta_1 \neq 0$.
  e p-value = .695 We cannot reject H_0: $\beta_0 = 0$. The population regression line should go through the origin.

11.15 a There is a very definite linear trend.
  b $b_0 = 4.6748$ $b_1 = 1.2272$
  c $t_{obs} = 10.44$ p-value = .000 The linear relationship is highly significant. The coefficient of determination is R-sq = 96.5%

d   sales = 4.6648 + 1.2272(6) = 12.028 ≈ 12

e   This model should not be used to predict sales for the month of January. Data for the month of January should be used to develop the model.

11.17 a   The dependent variable is sales.

b

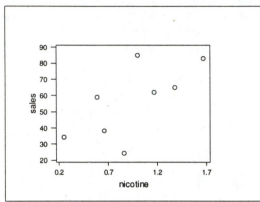

There is a weak upward linear trend.

c   The regression equation is sales = 24.9 + 33.1 nicotine

d t_{obs} = 2.24 p-value = 0.067 There is mildly significant evidence of a linear trend. The regression coefficient is positive (+33.1) therefore sales will increase when nicotine content increases

e   Sales increases 33.1 (x$100,000) for each additional milligram of nicotine.

11.19 a   The regression equation is Rating = 13.7 - 1.47 forty. This agrees with the answer in Exercise 11.7.

b t_{obs} = -1.45, p-value = .159 The null hypothesis should be accepted. This means that the proposed linear model is inadequate for explaining the relationship.

c   It seems as if the scouts should not be concerned with the time in the forty. A word of caution, however, there may be a relationship between the rating score and the time in the forty that is more complicated than a linear relationship.

11.21 a   There appears to be a downward linear trend.

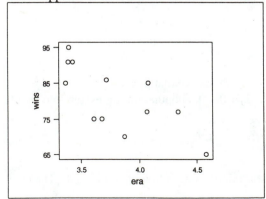

b   The regression equation is wins = 144 - 16.5 era. A unit change in era results in a decrease of 16.5 wins.
c t_{obs} = -3.24, p-value = .009 The null hypothesis should be rejected. The proposed linear model is highly significant.
d   This regression equation can be used to predict the number of wins based on the earned run average of the pitchers.

11.23 a   **Regression Analysis**

```
The regression equation is
Sat-M2 = 482 + 2.48 Profic2

20 cases used 12 cases contain missing values

Predictor       Coef        Stdev     t-ratio         p
Constant      481.54       14.35       33.57     0.000
Profic2       2.4823      0.7989        3.11     0.006

s = 24.80       R-sq = 34.9%      R-sq(adj) = 31.3%

Analysis of Variance

SOURCE         DF          SS          MS        F         p
Regression      1      5937.4      5937.4     9.65     0.006
Error          18     11071.2       615.1
Total          19     17008.6
```

R^2 = 34.9%; F_{obs} = 9.65, p-value = .006 The F statistic is highly significant.
b   The linear model found in Example 11.6 is slightly better than this model.
c

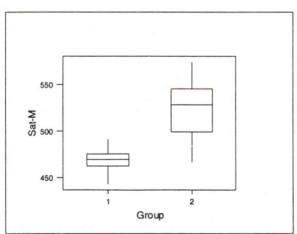

The SAT math scores for the second group are much more variable than in the first group. There is clearly a significant difference between the two medians.

11.25 a t_{obs} = 17.49, p-value = .000 The null hypothesis is rejected.
b R^2 = 88.4% and t_{obs} = 17.49.

c   One observation is unusual because its x value is so large (15.4). The other two are unusual because their standardized residual exceeds 2.0. In the scatterplot they are the two observations in the top right hand corner and the one observation below the main scatter.

Section 11.3 Estimation and Prediction

11.27 a $SSE = (n-2)s^2 = 22(9.25)^2 = 1882.375$
  b   For $x^* = 10$: $24.76 \pm 2.074*9.25\sqrt{(1/24 + 100/1280)}$
           $= 18.1$ to 31.4
        For $x^* = 15$: $28.11 \pm 2.074 *9.25\sqrt{(1/24 + 25/1280)}$
           $= 23.4$ to 32.9
        For $x^* = 20$: $31.46 \pm 2.074*9.25\sqrt{(1/24 + 0/1280)}$
           $= 27.5$ to 35.4
        For $x^* = 25$: $34.81 \pm 2.074*9.25\sqrt{(1/24 + 25/1280)}$
           $= 30.1$ to 39.6
        For $x^* = 30$: $38.16 \pm 2.074*9.25\sqrt{(1/24 + 100/1280)}$
           $= 31.5$ to 44.8

c
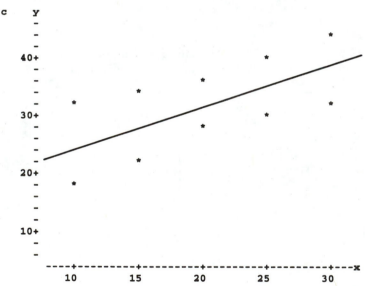

11.29 a   The t-ratio is highly significant and $R^2 = 84.4\%$ indicating that the regression line should fit the data.

  b

	Fit	Stdev.Fit	90.0% C.I.		90.0% P.I.	
For $x^*=10$	17.567	1.248	(15.304,	19.829)	(12.364,	22.769)
For $x^*=20$	22.467	0.817	(20.985,	23.948)	(17.554,	27.380)
For $x^*=25$	24.917	0.746	(23.564,	26.269)	(20.041,	29.792)
For $x^*=30$	27.367	0.817	(25.885,	28.848)	(22.454,	32.280)
For $x^*=40$	32.267	1.248	(30.004,	34.529)	(27.064,	37.469)

308

c   For $x^* = 25$ the confidence interval is the narrowest because $\bar{x} = 25$

11.31 a   The t-ratio is highly significant and $R^2 = 95.5\%$ indicating that the regression line fits the data very well.

b

	Fit	Stdev.Fit	95.0% C.I.		95.0% P.I.	
For x*=2	1.818	0.365	(0.924,	2.712)	(-0.010,	3.646)
For x*=4	3.091	0.286	(2.390,	3.792)	(1.350,	4.832)
For x*=6	4.364	0.237	(3.783,	4.944)	(2.667,	6.060)
For x*=8	5.636	0.237	(5.056,	6.217)	(3.940,	7.333)
For x*=10	6.909	0.286	(6.208,	7.610)	(5.168,	8.650)

11.33 a   Based on the scatterplot it appears that a straight line will fit the data.

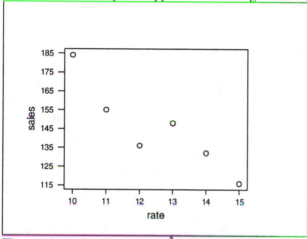

b   The t-ratio is significant and $R^2 = 82.8\%$ indicating that the regression line fits the data well.

c

	Fit	Stdev.Fit	98.0% C.I.		98.0% P.I.	
For x*=10	173.52	7.84	(144.15,	202.90)	(123.42,	223.62)
For x*=12	150.84	4.61	(133.57,	168.10)	(106.73,	194.94)
For x*=14	128.15	5.89	(106.10,	150.20)	(81.96,	174.34)
For x*=15	116.81	7.84	(87.44,	146.18)	(66.71,	166.91)

d   The confidence interval has the smallest length when $x^* = \bar{x} = 12.5$.

e   Because $x^* = 20$ is so far from $\bar{x} = 12.5$ the length of the interval will be much greater than it is when $x^* = 15$. This means that there could be a large amount of error in predicting the sales when the prime rate is 20%.

11.35 a and b

	Fit	Stdev.Fit	99.0% C.I.		99.0% P.I.	
For x* =1000	1686.2	35.1	(1580.4,	1792.0)	(1339.8,	2032.5)

11.37 a   The regression equation is   wins = 144 - 16.5 era
        The t-ratio is highly significant and $R^2 = 51.2\%$ indicating that that regression line fits the data very well.

b and c

	Fit	Stdev.Fit	99.0% C.I.		99.0% P.I.	
For x* =3.8	80.92	1.95	(74.74,	87.10)	(58.64,	103.20)

Section 11.4 Multiple Regression

11.39 a $\hat{y}$ = 5.6 - 3.1(5) + 4.9(.35) - 1.1(1.4) = -9.725

b   -3.1 ± 2.080*2.67 = -8.65 to 2.45

c t_{obs} = -1.1/.48 = -2.292   .01 < p-value < .025

     There is significant evidence that β_3 is less than 0.

11.41 $y = \beta_0 + \beta_1 x_1 + \beta_2 x_2 + \varepsilon$

where y = teacher's attitude, x_1 = years of experience, and x_2 = size of school

11.43 $y = \beta_0 + \beta_1 x_1 + \beta_2 x_2 + \varepsilon$

where y = family income, x_1 = education level, and x_2 = years residing in region

11.45 a   pulse = 58.3 + 5.82 dose

b R^2 = 67.8%

c t_{obs} = 5.825/1.072 = 5.43; p-value = .000

     There is highly significant evidence that dose should remain in the model.

d   The variable dose explains 67.85% of the variability in pulse and is highly significant. This is . a reasonable model to use to predict pulse.

11.47   The value of R^2 increases each time a variable is added to the model. With only 12 observations and seven predictor variables the value of R^2 should be rather large. So the increase from 51.2% to 94.7% is partly explained by the fact that the number of predictor variables is close to the sample size. The p-value for each t-ratio is computed with all other variables included in the model. For example the p-value (= .432) for batavg is computed with all other six variables in the model. We see from the following correlation matrix that the predictor variables are interrelated, and thus the variability explained by batavg is already explained by the other variables in the model. Consequently, its p-value comes up insignificant. The same is true for the other variables.

Correlations (Pearson)

	wins	batavg	rbi	stole	strkout	caught	errors
batavg	0.466						
rbi	0.647	0.588					
stole	-0.018	-0.221	-0.619				
strkout	-0.128	-0.504	-0.236	0.041			
caught	-0.117	-0.396	-0.641	0.835	0.533		
errors	-0.518	-0.388	-0.024	-0.334	-0.004	-0.205	
era	-0.715	0.010	-0.095	-0.451	0.084	-0.325	0.359

Notice also the era has the highest correlation with wins and is not correlated as much with the other independent variables. This explains why it is significant and the other variables are not. The variable rbi has the next highest correlation with wins and is not correlated with era. Entering rbi and era into the model will explain most of the variability explained by the other variables. We have

Regression Analysis

The regression equation is
wins = 76.9 + 0.0968 rbi - 15.2 era

Predictor	Coef	Stdev	t-ratio	p
Constant	76.94	18.54	4.15	0.002
rbi	0.09683	0.02139	4.53	0.000
era	-15.196	2.975	-5.11	0.000

s = 3.932   R-sq = 85.1%   R-sq(adj) = 81.8%

Analysis of Variance

SOURCE	DF	SS	MS	F	p
Regression	2	794.85	397.43	25.71	0.000
Error	9	139.15	15.46		
Total	11	934.00			

SOURCE	DF	SEQ SS
rbi	1	391.43
era	1	403.42

By adding rbi to the model containing era the value of R^2 increases from 51.2% to 85.1%, the F-statistic is highly significant, and both t-ratios are highly significant. This is a much improved model over the simple model containing only era and explains most of the variability that is explained by all seven variables.

11.49 a

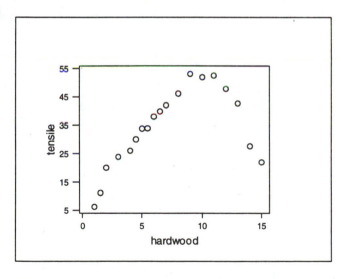

b   There is a curvilinear appearance to the scatterplot. A straight line will not fit the data.
c   R-sq = 30.5%, F = 7.47, p-value = 0.014 which is significant.
d   The answer to c does not agree with the answer to part b.

e   **Regression Analysis**

The regression equation is
tensile = - 6.67 + 11.8 hardwood - 0.635 hardsq

Predictor	Coef	Stdev	t-ratio	p
Constant	-6.674	3.400	-1.96	0.067
hardwood	11.764	1.003	11.73	0.000
hardsq	-0.63455	0.06179	-10.27	0.000

s = 4.420   R-sq = 90.9%   R-sq(adj) = 89.7%

Analysis of Variance

SOURCE	DF	SS	MS	F	p
Regression	2	3104.2	1552.1	79.43	0.000
Error	16	312.6	19.5		
Total	18	3416.9			

SOURCE	DF	SEQ SS
hardwood	1	1043.4
hardsq	1	2060.8

The value of R^2 increased from 30.5% to 90.9%, an enormous increase. The F-statistic increased 10-fold. The second degree polynomial model is an excellent model in comparison to the simple linear model.

11.51 a   **Regression Analysis**

The regression equation is
Rating = 2.53 + 0.834 Comp% + 4.09 AvgGn + 3.36 TD% - 4.19 INT%

Predictor	Coef	Stdev	t-ratio	p
Constant	2.5297	0.5315	4.76	0.000
Comp%	0.833568	0.009720	85.76	0.000
AvgGn	4.09479	0.06534	62.67	0.000
TD%	3.36220	0.03096	108.61	0.000
INT%	-4.18870	0.03687	-113.62	0.000

s = 0.1462   R-sq = 100.0%   R-sq(adj) = 100.0%

Analysis of Variance

SOURCE	DF	SS	MS	F	p
Regression	4	3746.96	936.74	43799.82	0.000
Error	25	0.53	0.02		
Total	29	3747.49			

SOURCE	DF	SEQ SS
Comp%	1	2777.96
AvgGn	1	296.49
TD%	1	396.44
INT%	1	276.07

Unusual Observations

Obs.	Comp%	Rating	Fit	Stdev.Fit	Residual	St.Resid
10	61.5	87.800	87.458	0.037	0.342	2.42R

R denotes an obs. with a large st. resid.

b $R^2 = 100\%$ with all four variables in the model. It is unusual to obtain a value of R^2 that explains 100% of the variability. This multiple regression model must be equivalent to the model used by the NFL to compute the ratings of the quarterbacks.

c   All t-ratios are highly significant. Based on these t-ratios no variables should be removed from the model.

d   R-sq = 87.1%. 87.1% of the variability in the Rating is explained by these two variables whereas 100% is explained by all four variables above. If one variable is difficult to measure we might think of removing it from the model, but because data for all four variables are easily obtained we should keep all variables in the model.

```
e           Fit   Stdev.Fit       95.0% C.I.              95.0% P.I.
Aikman   108.557       0.148   ( 108.252, 108.863)   ( 108.128, 108.987) XX
Odonnell  51.282       0.156   (  50.961,  51.603)   (  50.842,  51.723) XX
```

Aikman was definitely the better quarterback in Superbowl XXX.

Section 11.5 Checking Model Adequacy

11.53 It appears that the residuals are decreasing as x increases, which indicates that the variance is not constant with respect to x.

11.55 The standardized residuals increase as the fits increase. This indicates that the variance is not constant with respect to x.

11.57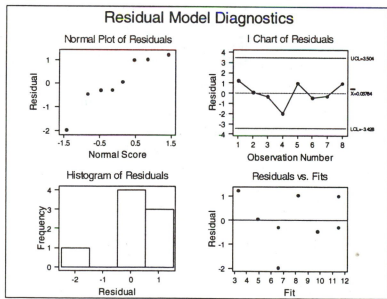

There may be a slight departure from normality but otherwise there is no unusual behavior shown in this residual analysis

11.59 a The scatterplot indicates a fairly strong linear relationship between the number of library books checked out and the verbal test scores of the eighth graders.
b R^2 = 85.2% and the t-ratio is highly significant (p-value = .000)
c The standardized residual show a definite curvilinear pattern. We should consider a second degree polynomial model.

11.61 a The residuals are close to normally distributed.
b The Residuals vs. Fits plot shows a more random pattern than before.
c This model appears to be an improvement over the simple linear model.

11.63 **Regression Analysis**

```
The regression equation is
output = 2.98 - 6.93 recipvel

Predictor      Coef       Stdev      t-ratio        p
Constant    2.97886     0.04490       66.34     0.000
recipvel    -6.9345      0.2064      -33.59     0.000

s = 0.09417     R-sq = 98.0%      R-sq(adj) = 97.9%

Analysis of Variance

SOURCE        DF          SS          MS         F        p
Regression     1      10.007      10.007   1128.43    0.000
Error         23       0.204       0.009
Total         24      10.211

Unusual Observations
Obs. recipvel      output       Fit   Stdev.Fit    Residual    St.Resid
  1      0.408     0.1230    0.1484      0.0474     -0.0254     -0.31 X
 11      0.183     1.5010    1.7065      0.0191     -0.2055     -2.23R
 16      0.143     1.8000    1.9882      0.0219     -0.1882     -2.06R
```

R denotes an obs. with a large st. resid.
X denotes an obs. whose X value gives it large influence.
a The new R^2 is 98.0%. Very much improved over the previous model.
b The new F-statistic is 1128.43, again much improved over the previous model.
c The new residual plot shows a random pattern.

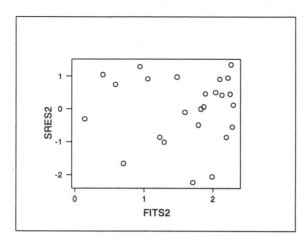

11.65 a The regression equation is   sales = 55.0 - 6.82 rd
rd: t_{obs} = -2.60, p-value = 0.027 The regression coefficient is significantly different from 0.

b R-sq = 40.3% which is the percent of variability in sales explained by R & D.

c F = 6.74, p-value = 0.027. The p-value for the F statistic is the same as the p-value for the t-statistic because the two test are equivalent in the simple linear regression model. More specifically, $F = t^2$.

d Unusual Observations

Obs.	rd	sales	Fit	Stdev.Fit	Residual	St.Resid
6	6.30	41.00	12.07	8.05	28.93	2.85R

R denotes an obs. with a large st. resid.
The standardized residual for observation 6 is unusually large.

e **Regression Analysis**

The regression equation is
sales = 69.9 - 12.1 rd

11 cases used 1 cases contain missing values

Predictor	Coef	Stdev	t-ratio	p
Constant	69.945	5.193	13.47	0.000
rd	-12.082	1.462	-8.26	0.000

s = 5.901   R-sq = 88.4%   R-sq(adj) = 87.1%
Analysis of Variance

SOURCE	DF	SS	MS	F	p
Regression	1	2377.1	2377.1	68.26	0.000
Error	9	313.4	34.8		
Total	10	2690.5			

f R^2 improves from 40.3% to 88.4%. The F-statistic and t-statistic are highly significant. There are no unusual observations.

g The slope of the regression line found in part a is almost half the slope of the resistant line, whereas the regression line found in part e is almost identical to the resistant line.

Section 11.6 Summary and Review

11.67 a   Independent: population density Dependent: robbery rate
        b   Independent: attitude score Dependent: achievement score
        c   Independent: number of fish Dependent: growth rate
        d   Independent: expenditures per student Dependent: teacher's salary

11.69 a   Slope = -2/3, y-intercept = 2
        b   Slope = 7.4/3.1, y-intercept = -12/3.1
        c   Slope = -5.6/4.1, y-intercept = -10/4.1

11.71 a

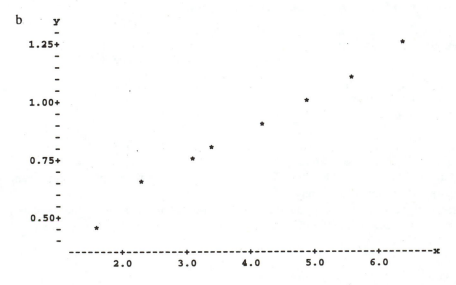

The relationship between x and y appears to be curvilinear.

b

316

There appears to be a strong linear relationship between x and y.

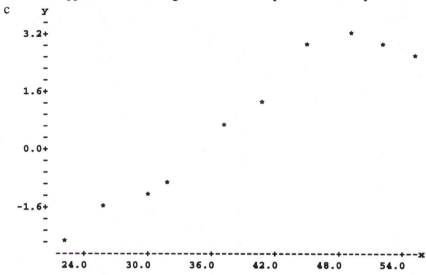

The relationship between x and y appears to be curvilinear.

11.73 a x' = 100, $\hat{y}$ = 7.3; x' = 160 $\hat{y}$ = 2.86

b x' = 1.5, $\hat{y}$ = .502; x' = 5.0, $\hat{y}$ = 1.02

c x' = 30, $\hat{y}$ = -.96; x' = 60, $\hat{y}$ = 4.53

11.75

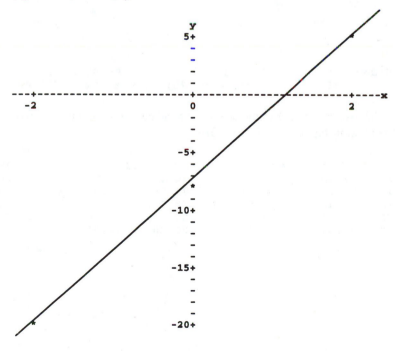

$x = 5,\ \hat{y} = 23.7,\ e = -5.7$

$x = 5,\ \hat{y} = 23.7,\ e = -3.7$

11.77

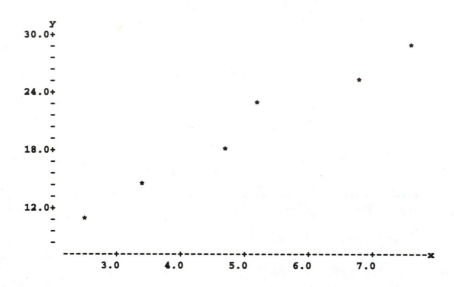

Overall, there appears to be a strong linear relationship between x and y.

11.79 a $\mu_y = 200 - .35(45) = 184.25$

     b $184.25 \pm 2(1.5)$ The interval is $(181.25, 187.25)$.

     c $-.35$

11.81

Fit	Stdev.Fit	95.0% C.I.		95.0% P.I.	
56.40	1.89	(52.05,	60.75)	(42.72,	70.08)

11.83 Because of the bivariate outlier in the upper left hand corner a resistant line would better describe the relationship between IQ and GPA.

11.85

	Fit	Stdev.Fit	95.0% C.I.		95.0% P.I.	
For x'=10	71.85	2.53	(66.45,	77.25)	(50.80,	92.90)
For x'=15	78.01	2.31	(73.08,	82.95)	(57.08,	98.95)
For x'=20	84.18	2.54	(78.76,	89.60)	(63.13,	105.23)
For x'=25	90.34	3.12	(83.69,	97.00)	(68.94,	111.75)
For x'=30	96.51	3.90	(88.20,	104.81)	(74.53,	118.48)

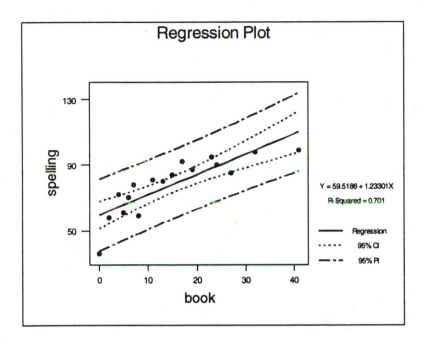

Regression Plot

$$Y = 59.5186 + 1.23301X$$

R-Squared = 0.701

— Regression
..... 95% CI
—·— 95% PI

11.87 **Regression Analysis**

The regression equation is
confid = 85.5 - 4.23 educat

Predictor	Coef	Stdev	t-ratio	p
Constant	85.53	12.51	6.84	0.000
educat	-4.231	1.020	-4.15	0.001

s = 10.35   R-sq = 48.9%   R-sq(adj) = 46.0%

Analysis of Variance

SOURCE	DF	SS	MS	F	p
Regression	1	1842.7	1842.7	17.20	0.001
Error	18	1928.3	107.1		
Total	19	3770.9			

Unusual Observations

Obs.	educat	confid	Fit	Stdev.Fit	Residual	St.Resid
8	12.0	62.00	34.76	2.31	27.24	2.70R

R denotes an obs. with a large st. resid.

a

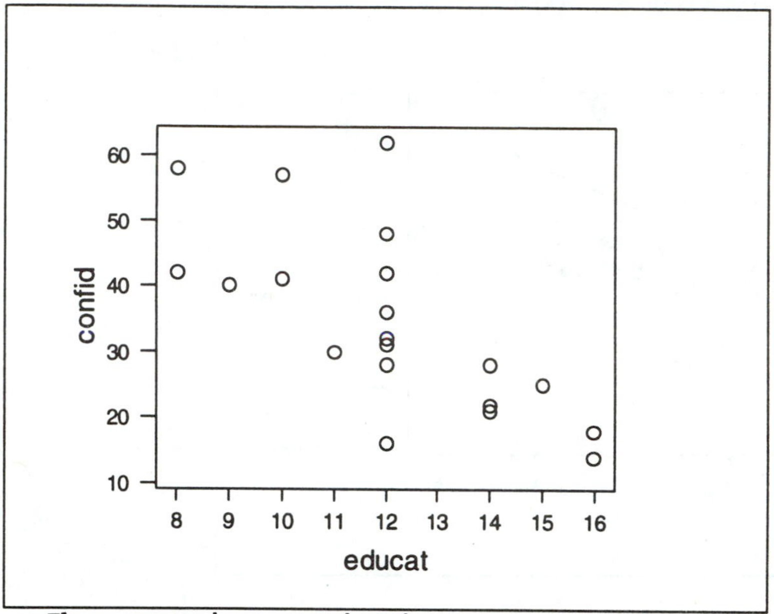

The regression equation is **confid = 85.5 - 4.23 educat**

b SSE = **1928.3** **s = 10.35**

c

	Predictor	Coef	Stdev	t-ratio	p	
	educat	-4.231	1.020	-4.15	0.001001	Reject H_0:β_1=0

d **R-sq = 48.9%** Thus, 48.9% of the variability in confidence is explained by education.

e This model adequately describes the relationship between the two variables.

11.89

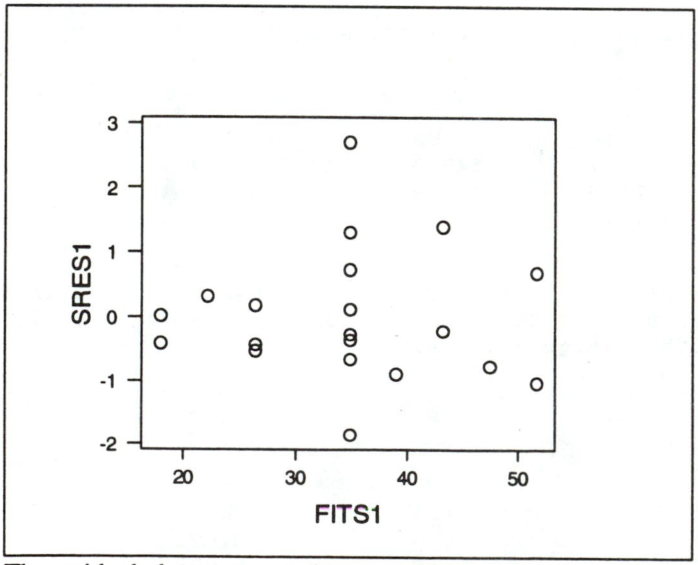

The residual plot appears to have a random pattern.

11.91 **Regression Analysis**

The regression equation is
sales = 3246 + 49.9 months

Predictor	Coef	Stdev	t-ratio	p
Constant	3246.4	574.5	5.65	0.000
months	49.85	29.89	1.67	0.134

s = 977.6   R-sq = 25.8%   R-sq(adj) = 16.5%

Analysis of Variance

SOURCE	DF	SS	MS	F	p
Regression	1	2658223	2658223	2.78	0.134
Error	8	7645216	955652		
Total	9	10303440			

a   The regression equation is   sales = 3246 + 49.9 months

b   R-sq = 25.8%

c
Predictor	Coef	Stdev	t-ratio	p
months	49.85	29.89	1.67	0.134

d

Residual Model Diagnostics

A possible violation of assumptions is that the variance does not appear to be constant.

11.93 a   **Regression Analysis**

The regression equation is
price = 33.4 - 0.393 harvest

Predictor	Coef	Stdev	t-ratio	p
Constant	33.376	4.097	8.15	0.001
harvest	-0.39267	0.06303	-6.23	0.003

```
s = 1.015        R-sq = 90.7%      R-sq(adj) = 88.3%
```

Analysis of Variance

SOURCE	DF	SS	MS	F	p
Regression	1	39.987	39.987	38.81	0.003
Error	4	4.121	1.030		
Total	5	44.108			

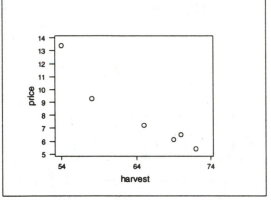

b **Predictor   Coef   Stdev   t-ratio   p**

harvest   -0.39267   0.06303   -6.23   0.003 Reject $H_0: \beta_1 = 0$

c **R-sq = 90.7%** This model explains 90.7% of the variability.

d The model seems adequate.

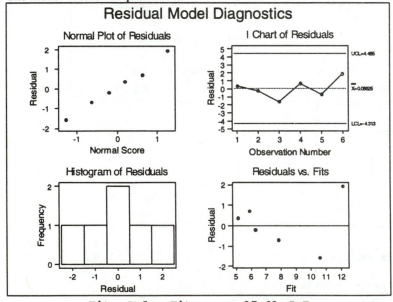

e
```
            Fit   Stdev.Fit      95.0% C.I.              95.0% P.I.
For x*=60  9.816   0.508    (   8.405,  11.227)   (   6.663,  12.968)
For x*=30 21.596   2.224    (  15.419,  27.773)   (  14.807,  28.385) XX
X  denotes a row with X values away from the center
XX denotes a row with very extreme X values
```

The prediction for $x^* = 60$ is more reliable because 60 is within the range of values of the collected data. $x^* = 30$ is far removed from the rest of the data. The resulting estimate would be and extrapolation beyond the collected data.

11.95 a **Correlations (Pearson)**

	year	total	probat	jail	prison
total	0.998				
probat	0.990	0.992			
jail	0.985	0.989	0.968		
prison	0.990	0.989	0.963	0.990	
parole	0.967	0.969	0.932	0.981	0.990

All the variables are highly correlated with year.

b

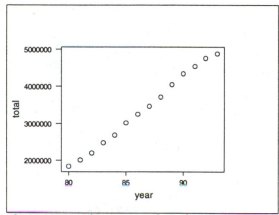

The regression equation is
total = -18147248 + 248780 year

Predictor	Coef	Stdev	t-ratio	p
Constant	-18147248	360146	-50.39	0.000
year	248780	4159	59.82	0.000

s = 62731   **R-sq = 99.7%**   **R-sq(adj) = 99.6%**

Analysis of Variance

SOURCE	DF	SS	MS	F	p
Regression	1	1.40803E+13	1.40803E+13	3578.06	0.000
Error	12	47222210560	3935184128		
Total	13	1.41275E+13			

Unusual Observations

Obs.	year	total	Fit	Stdev.Fit	Residual	St.Resid
14	93.0	4879600	4989291	31810	-109691	-2.03R

R denotes an obs. with a large st. resid.
99.7% of the variability in number of prisoners is explained by year. There is no need to consider other variables in the model.

c The residual pattern appears random.

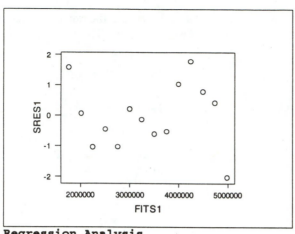

d Regression Analysis

```
The regression equation is
parole = - 2881080 + 37834 year

Predictor        Coef        Stdev      t-ratio         p
Constant     -2881080       250491       -11.50     0.000
year            37834         2893        13.08     0.000

s = 43631        R-sq = 93.4%      R-sq(adj) = 92.9%

Analysis of Variance

SOURCE       DF            SS            MS          F         p
Regression    1 3.25646E+11 3.25646E+11     171.06     0.000
Error        12 22844065792  1903672192
Total        13 3.48490E+11
```
Year explains 93.4% of the variability in parole.

e The residual plot shows a curvilinear pattern. The model needs to be revised. A polynomial model would be a good prospect.

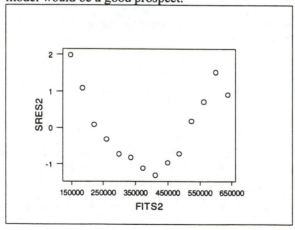

11.97 a

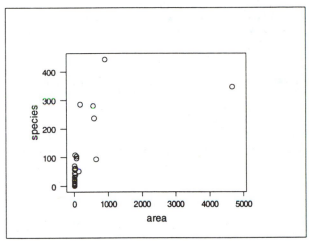

b The regression equation is
species = 63.8 + 0.0820 area

Predictor	Coef	Stdev	t-ratio	p
Constant	63.78	17.52	3.64	0.001
area	0.08196	0.01971	4.16	0.000

s = 91.73   R-sq = 38.2%   R-sq(adj) = 36.0%

Analysis of Variance

SOURCE	DF	SS	MS	F	p
Regression	1	145470	145470	17.29	0.000
Error	28	235611	8415		
Total	29	381081			

Unusual Observations

Obs.	area	species	Fit	Stdev.Fit	Residual	St.Resid
16	4669	347.0	446.5	88.5	-99.5	-4.11RX
25	904	444.0	137.9	21.0	306.1	3.43R
27	171	285.0	77.8	16.8	207.2	2.30R

R denotes an obs. with a large st. resid.
X denotes an obs. whose X value gives it large influence.
$R^2 = 38.2\%$, F = 17.29, p-value = .000 (highly significant) Additional variables may
increase R^2 and improve the model.

c A log transformation or square root transformation may help explain the relationship.

d The regression equation is
species = 45.3 + 25.7 lnarea

Predictor	Coef	Stdev	t-ratio	p
Constant	45.29	14.48	3.13	0.004
lnarea	25.699	3.835	6.70	0.000

s = 72.30   R-sq = 61.6%   R-sq(adj) = 60.2%

325

Analysis of Variance

SOURCE	DF	SS	MS	F	p
Regression	1	234723	234723	44.91	0.000
Error	28	146358	5227		
Total	29	381081			

Unusual Observations

Obs.	lnarea	species	Fit	Stdev.Fit	Residual	St.Resid
25	6.81	444.0	220.2	24.1	223.8	3.28R

R denotes an obs. with a large st. resid.

The value of $R^2 = 61.6\%$ improved substantially, F= 44.91, p-value =.000 is highly significant. This model appears to be an improvement over the previous model.

12

Analysis of Variance

Section 12.2 Comparison of Several Means: The One-Way ANOVA

12.1 a   The stem and leaf plots suggest that the variances are homogeneous.
     b   Based on the stem and leaf plots there are no serious departures from normality.
     c   The standard deviations of the three samples are very close indicating that the variances are homogeneous.
     d $MSE = s_p^2 = [17(15.12)^2 + 16(12.39)^2 + 15(12.44)^2]/48 = 180.4988$
     e   The grand mean cannot be calculated by averaging the three means because the group sizes are different.

12.3 a $SSE = 8662$
     b $MSE = 180$
     c $SST = 1069$
     d $MST = 534$
     e   The square of the pooled standard deviation is the mean square error, MSE.
     f $H_0: \mu_1 = \mu_2 = \mu_3$ versus H_a: at least 2 μ_i's differ.
     g $F = 2.96$, p-value = .061 The null hypothesis can be rejected but the results are only mildly significant.

12.5 a $H_0: \mu_1 = \mu_2 = \mu_3$ versus H_a: at least 2 μ_i's differ.
     b $F = 6.05$, p-value = .004 The null hypothesis should be rejected; the results are highly significant.

12.7 a $T = 2.10$, p-value =.047 and $F = 4.43$, p-value = .047 The relationship between the two is $T^2 = F$ because there are only two groups. Because of the exact relationship between the two their p-values are exactly the same.
     b   The pooled standard deviations for the two test are exactly the same (except for slight round off error) Furthermore, $MSE = 91.2 = s_p^2 = (9.55)^2$.
     c   The independent pooled t-test and the F-test are equivalent when we have only two groups to compare.

12.9 a   Based on the boxplots it is doubtful that there is a significant difference between the thickness of the wafers.
     b   The measurements are not independent because four measurements of the thickness of the oxide layer are taken on the same wafer.

12.11 The largest standard deviation is less than twice the smallest standard deviation so we can conclude that the variances are homogeneous. Based on the p-value the null hypothesis should be rejected. There is a highly significant difference between the number of tasks completed in a 10 minute period by subjects treated with different combinations of a drug and electroshock.

12.13 Condition A: $14.250 \pm 2.048(1.854/\sqrt{8})$ The confidence interval is (12.91, 15.59)
Condition B: $19.125 \pm 2.048(1.854/\sqrt{8})$ The confidence interval is (17.78, 20.47)
Condition C: $11.000 \pm 2.048(1.854/\sqrt{8})$ The confidence interval is (9.66, 12.34)
Condition D: $16.625 \pm 2.048(1.854/\sqrt{8})$ The confidence interval is (15.28, 17.97)

12.15 a $H_0: \mu_1 = \mu_2 = \mu_3$ versus H_a: at least 2 μ_i's differ.
   b There does not appear to be any violations of the normality assumption. All indications are that the variances are homogeneous. If the samples are random and independent the assumptions for an F-test are satisfied.
   c F = 12.17, p-value = .000 The null hypothesis is rejected. There is a highly significant difference between the average weight gained under the three rations.

12.17 a The normality assumption may be in question.
   b None of the samples fail the normality test.

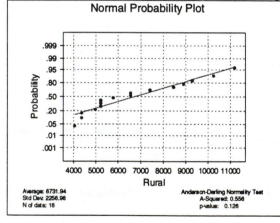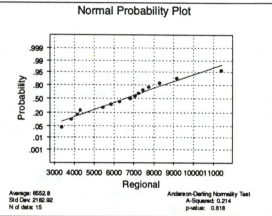

328

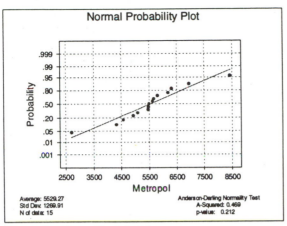

Normal Probability Plot

Average: 5529.27
Std Dev: 1269.91
N of data: 15

Anderson-Darling Normality Test
A-Squared: 0.469
p-value: 0.212

c   **Descriptive Statistics**

Variable	N	Mean	Median	TrMean	StDev	SEMean
Rural	16	6732	6158	6607	2257	564
Regional	15	6553	6801	6428	2183	564
Metropol	15	5529	5493	5524	1270	328

Variable	Min	Max	Q1	Q3
Rural	4027	11191	5035	8784
Regional	3339	11391	4288	7733
Metropol	2696	8435	4925	6177

Based on the standard deviations no variance is more than four times as large as any other variance. The homogeneous variances assumption is satisfied.

d H_0: $\mu_1 = \mu_2 = \mu_3$ versus H_a: at least 2 μ_i's differ.

e   **One-Way Analysis of Variance**

Analysis of Variance

Source	DF	SS	MS	F	p
Factor	2	12838182	6419091	1.67	0.201
Error	43	165698288	3853448		
Total	45	178536464			

Individual 95% CIs For Mean
Based on Pooled StDev

Level	N	Mean	StDev	
Rural	16	6732	2257	(----------*---------)
Regional	15	6553	2183	(----------*---------)
Metropol	15	5529	1270	(---------*----------)

```
                                  -----+---------+---------+---------+-
                                     5000      6000      7000      8000
```

Pooled StDev = 1963

Based on the p-value the differences between the costs of laminectomies are insignificant.

Section 12.3 Tukey's Multiple Comparison Procedure

12.19 a 5.99   b 3.70   c 5.76

12.21 a **ANOVA table**

Source	SS	df	MS	F	p-value
Diet	228.55	3	76.183	14.494	< .001
Error	84.1	16	5.256		

b Based on the p-value the difference between the treatment means is highly significant.

c $W = 4.05\sqrt{5.256/5} = 4.1525$

$\mu_C \quad \mu_B \quad \mu_A \quad \mu_D$

The diet plans B, A and D do not differ significantly. Diet plan C is significantly better than the other three diet plans.

12.23 $W_{12} = 3.96\sqrt{1.75/2}\sqrt{1/5 + 1/7} = 2.17$
$W_{13} = 3.96\sqrt{1.75/2}\sqrt{1/5 + 1/7} = 2.17$
$W_{14} = 3.96\sqrt{1.75/2}\sqrt{1/5 + 1/6} = 2.24$
$W_{23} = 3.96\sqrt{1.75/2}\sqrt{1/7 + 1/7} = 1.98$
$W_{24} = 3.96\sqrt{1.75/2}\sqrt{1/7 + 1/6} = 2.06$
$W_{34} = 3.96\sqrt{1.75/2}\sqrt{1/7 + 1/6} = 2.06$

$\mu_3 \quad \mu_2 \quad \mu_1 \quad \mu_4$

The mean reaction effect from drug 3 is greater than all the other means.

12.25 **Tukey's pairwise comparisons**

Family error rate = 0.0500
Individual error rate = 0.0104

Critical value = 3.74

Intervals for (column level mean) - (row level mean)

	1	2	3
2	-1.741		
	0.741		
3	0.072	0.572	
	2.553	3.053	
4	-2.553	-2.053	-3.866
	-0.072	0.428	-1.384

$\mu_4 \quad \mu_2 \quad \mu_1 \quad \mu_3$

The mean from group 3 (No drug w/shock) is significantly below the means for the other groups.

12.27 **Tukey's pairwise comparisons**

      Family error rate = 0.0500
  Individual error rate = 0.0106

Critical value = 3.79

Intervals for (column level mean) - (row level mean)

	1	2	3
2	-2.772 0.044		
3	-3.681 -0.865	-2.317 0.499	
4	-1.953 0.863	-0.590 2.226	0.319 3.135

μ_3 μ_2 μ_4 μ_1

The mean test score for those students using method III is significantly greater than those using methods I and IV

12.29 Based on the F-statistic and its p-value there is a highly significant difference between the mean residual contaminant left by the three cleansing agents. All of the confidence intervals under Tukey's pairwise comparison show a significant difference between the corresponding means. In other words, the mean for cleansing agent A is significant below the other two and the mean for cleansing agent B is significantly above the other two cleansing agents.

Section 12.4 The Kruskal-Wallis Test

12.31 The Kruskal-Wallis test is a nonparametric test that is an alternative to the ordinary F-test that is used to compare several population means. To execute the rank-transformation form of the test ranks are collectively assigned to the combined samples and then the ordinary F-test is applied to the ranks.

12.33 $SST = 10(12.45 - 15.5)^2 + 10(14.7 - 15.5)^2 + 10(19.35 - 15.5)^2 = 247.65$
$SSE = 9(8.67)^2 + 9(8.43)^2 + 9(8.72)^2 = 2000.45$

Source	df	SS	MS	F	p-value
Treatment	2	247.65	123.825	1.67	> .10
Error	27	2000.45	74.1		

12.35 a   All three distributions are skewed right
     b   The Kruskal-Wallis should be conducted on these data.

12.37 a Based on the boxplots there is no severe skewness or extreme outliers in any of the three groups. The population variances seem homogeneous. There is no serious departure from normality in any of the three distributions.

b There are no serious departures from assumptions for the ordinary F-test. Obviously the assumptions for the Kruskal-Wallis are also satisfied. It is recommended that these data be analyzed with the ordinary F-test.

12.39 a All three distributions are skewed right, in which case, normality is not possible. The distributions, however, are similar in shape.

b Because the distributions are similar in shape, the Kruskal-Wallis test can be applied to the data.

12.41 If the distributions are long-tailed the Kruskal-Wallis test is the proper procedure to use to compare the centers of the distributions. The ordinary F-test is sensitive to outliers and the K-W test is robust.

12.43 a

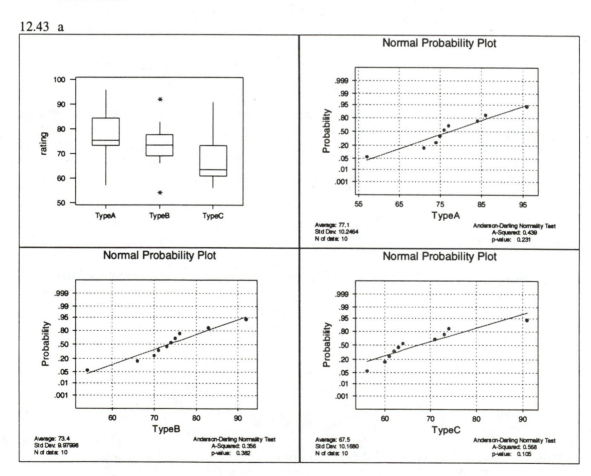

There are outliers in the type B group and skewness in the other two distributions.

b Because normality is questionable we should used the Kruskal-Wallis test.

c   **One-Way Analysis of Variance**

Analysis of Variance on Ranks

Source	DF	SS	MS	F	p
Type	2	484.0	242.0	3.72	0.037
Error	27	1756.6	65.1		
Total	29	2240.5			

Individual 95% CIs For Mean
Based on Pooled StDev

Level	N	Mean	StDev				
1	10	20.150	7.951			(--------*-------)	
2	10	16.000	8.416		(--------*-------)		
3	10	10.350	7.818	(-------*--------)			

```
                                     --+---------+---------+---------+----
Pooled StDev =    8.066              6.0      12.0      18.0      24.0
```

Based on these results there is a significant difference between the efficiency ratings for the three types of oil heaters.

12.45 a   The outlier in the third state is so extreme that it is difficult to compare the three boxplots.

    b   Because of the extreme outlier the ordinary F-test should not be used to analyze these data.

    c   This is a situation where one observation distorts the data so much that it should probably be removed before analyzing the data.

    d   After removing the outlier we see that the three distributions are skewed right.

    e   Because of the skewness the Kruskal-Wallis test should be used to compare the medians.

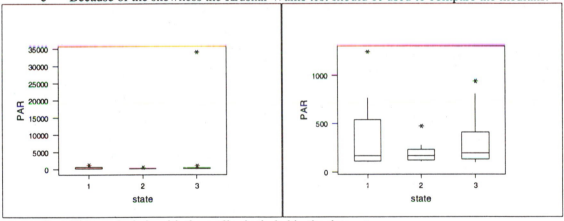

    f   This analysis is with the outlier included in the data set:

One-Way Analysis of Variance

Analysis of Variance on ranks

Source	DF	SS	MS	F	p
state	2	611	305	0.85	0.432
Error	62	22269	359		
Total	64	22880			

Individual 95% CIs For Mean
Based on Pooled StDev

Level	N	Mean	StDev				
1	21	31.95	22.34		(-----------*----------)		
2	15	28.47	14.53	(-------------*-------------)			
3	29	36.10	18.25			(--------*---------)	

```
                                    ----+---------+---------+---------+--
Pooled StDev =    18.95              21.0      28.0      35.0      42.0
```

The following analysis is without the one extreme outlier:
One-Way Analysis of Variance

```
Analysis of Variance on ranks
Source      DF        SS           MS          F         p
state        2       435          218       0.62     0.541
Error       61     21405          351
Total       63     21840
```

```
                                  Individual 95% CIs For Mean
                                  Based on Pooled StDev
  Level      N       Mean       StDev   ----+---------+---------+---------+--
    1       21      31.95       22.34               (-----------*----------)
    2       15      28.47       14.53   (-------------*------------)
    3       28      35.07       17.71                    (---------*---------)
                                        ----+---------+---------+---------+--
Pooled StDev =      18.73                 21.0      28.0      35.0      42.0
```

Excluding the extreme outlier does not have much effect on the analysis of the ranks. As far as the ranks are concerned it is just the 65th largest observation. Excluding it means that the largest observation is 64 instead of 65. Its effect is minimal. For the ordinary F-test, however, it would have an enormous effect on the analysis.

g   Based on the analysis in part f there is an insignificant difference between the median Problem Asset Ratios for the three states.

Section 12.5 Summary and Review

12.47 The ordinary F-test.

12.49 a   3.01, 3.72, 4.72
       b   2.53, 3.03, 3.70
       c   2.25, 2.62, 3.12

12.51 a   5.29   b 3.67   c 5.39

12.53

Source	SS	df	MS	F	p-value
Treatment	2313.04	2	1156.52	6.38	< .01
Error	8154.10	45	181.2		
Total	10467.14	47			

12.55 a   There are five treatment levels.
       b

Source	SS	df	MS	F	p-value
Treatment	428	4	107	2.903	.025 < p-value < .05
Error	1032	28	36.857		
Total	1460	32			

c   There is a significant difference between the treatment levels.

12.57

Source	DF	SS	MS	F	p
Factor	2	573	286	1.61	0.212
Error	41	7291	178		
Total	43	7864			

12.59 a   The distributions are symmetric. Normality seems reasonable in all three distributions.

b   Treat1: midrange = 31.5, midQ = 31.625, mean = 32.12, median = 32.5, trimmed = 32.12
Treat2: midrange = 40, midQ = 40.375, mean = 40.12, median = 40, trimmed = 40.12
Treat3: midrange = 38, midQ = 37.25, mean = 37.25, median = 37, trimmed = 37.25
In all three cases the midsummary statistics are very close in value suggesting that the distributions are symmetric.

c   The variance of group 1 appears larger than the variances of the other two groups but not enough to suggest that they are not homogeneous. The standard deviations do not rule out homogeneous variances.

d   The assumptions for the ordinary F-test are satisfied. If the assumptions for the F-test are satisfied then the assumptions for the Kruskal-Wallis test are automatically satisfied.

e   Based on the boxplots it appears that there is a significant evidence to reject the hypothesis that the medians of the three treatments are the same.

12.61 a   There is probably not a significant difference between the median ages for adjacent phases but examining the medians for phase 1 and phase 4 we see a difference that is significant. There is no unusual behavior in the boxplots that would suggest that the assumptions for the ordinary F-test are not satisfied.

b   Based on the p-value = .000 ($< .005$) there is highly significant evidence that the null hypothesis should be rejected. Phases 4 and 3 are nonoverlapping, phases 3 and 1 are nonoverlapping, but phases 3 and 2 appear to overlap. It is difficult to assess whether or not phases 2 and 1 overlap.

c   Based on the intervals produced by Tukey's multiple comparison μ_1 and μ_2 are not statistically different and μ_2 and μ_3 are not statistically different. This suggest that phases 1 and 2 overlap and phases 2 and 3 overlap.

12.63 Recall that there was one extreme outlier in the Texas data. This one outlier distorts both the mean and standard deviation of the sample. The mean of 1452 is not a realistic measure of the center of the distribution. The standard deviation of 6270 is not a realistic measure of the variability of the distribution. Moreover, it causes the MSE to be unusually large which in turn causes the F-statistic to be unusually small (MSE is in the denominator). Consequently, the mean for Texas is inflated but the F-test does not detect it because of the unusually large MSE. It is clear that we have violated two basic assumptions for the F-test. First, the distribution is not normally distributed and second the variances are not homogeneous. It is for this reason that we recommended removing the outlier and analyzing with the Kruskal-Wallis test.

12.65 a $H_0: \mu_1 = \mu_2 = \mu_3$ versus H_a: at least 2 μ_i's differ.

b   Based on the standard deviations the variances are homogeneous. We need normal probability plots to check the normality assumption, but with so few observations it would be difficult to detect departures from normality.

c   **One-Way Analysis of Variance**

```
Analysis of Variance
Source      DF        SS        MS         F        p
Factor       2    0.4000    0.2000      6.32    0.013
Error       12    0.3800    0.0317
Total       14    0.7800
                                    Individual 95% CIs For Mean
                                    Based on Pooled StDev
Level       N      Mean     StDev   -+---------+---------+---------+-----
1%           5    1.9000    0.1581  (------*------)
5%           5    2.1000    0.1871           (------*------)
10%          5    2.3000    0.1871                    (------*------)
                                    -+---------+---------+---------+-----
Pooled StDev =    0.1780             1.75      2.00      2.25      2.50
```

d   Based on the F-test there is a significant difference between the mean dissolving times for the different levels of impurity. In other words, the level of impurity has a significant effect on the solubility of the aspirin tablet.

12.67 a   The distributions look reasonably symmetric without unusually long tails.

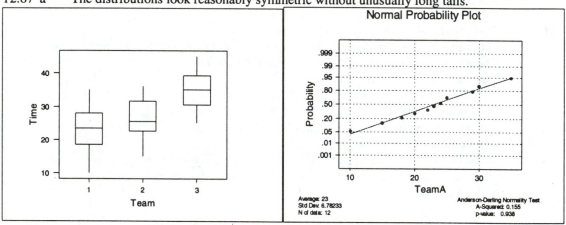

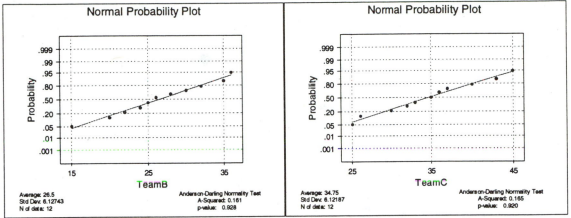

b   The normal probability plots show no significant departures from normality.

c   The assumptions for the F-test are satisfied and thus it should be used to compare the centers of the distributions.

12.69 a   Based on the normal probability plots and the appearance of the side-by-side boxplots we classify the four distributions as skewed right.

b   Because the normality assumption is in doubt and the fact that the distributions are similarly skewed we should use the Kruskal-Wallis test to compare the centers of the distributions.

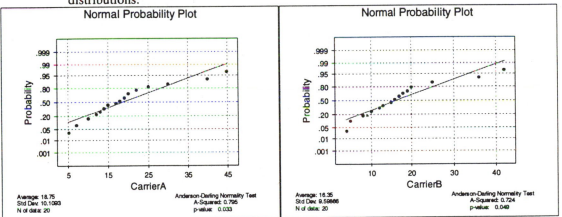

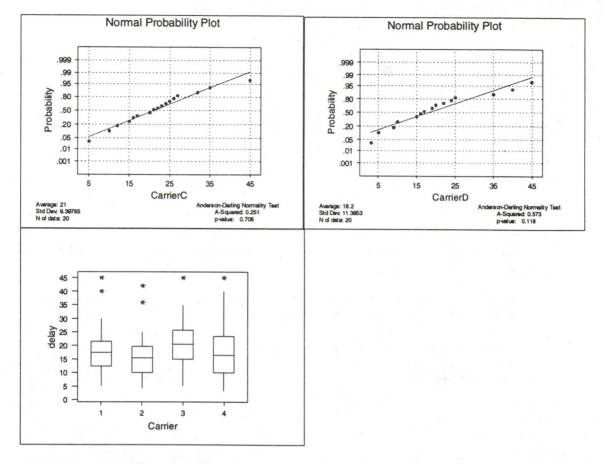

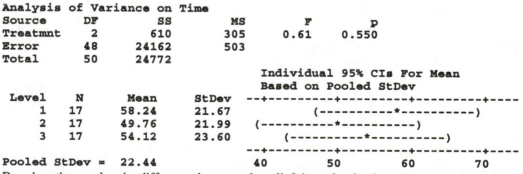

12.71 Because the difference is insignificant there is no need to conduct a multiple comparison test.

12.73 `One-Way Analysis of Variance`

```
Analysis of Variance on Time
Source      DF        SS        MS        F         p
Treatmnt     2       610       305      0.61     0.550
Error       48     24162       503
Total       50     24772
                                    Individual 95% CIs For Mean
                                    Based on Pooled StDev
Level       N      Mean      StDev  --+---------+---------+---------+----
    1      17     58.24     21.67                 (----------*----------)
    2      17     49.76     21.99      (----------*----------)
    3      17     54.12     23.60         (----------*----------)
                                    --+---------+---------+---------+----
Pooled StDev =   22.44            40        50        60        70
```
Based on the p-value the difference between the relief times for the three treatments for arthritis is insignificant.

338

12.75 a

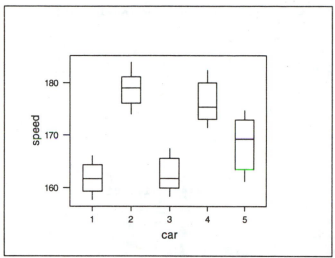

There are no indications in these boxplots that would suggest that we should not conduct an ordinary F-test. The distributions are symmetric and the variances are homogeneous. Based on the boxplots it appears that the normality assumption is satisfied. At least there are no outliers or severe skewness.

b H_0: $\mu_1 = \mu_2 = \mu_3 = \mu_4 = \mu_5$ versus H_a: at least 2 μ_i's differ.

One-Way Analysis of Variance

Analysis of Variance on speed

Source	DF	SS	MS	F	p
car	4	1456.5	364.1	25.15	0.000
Error	25	362.0	14.5		
Total	29	1818.5			

```
                                Individual 95% CIs For Mean
                                Based on Pooled StDev
Level    N     Mean     StDev   ----+---------+---------+---------+--
  1      6    161.72    2.92    (----*----)
  2      6    178.78    3.38                              (---*----)
  3      6    162.40    3.36    (----*----)
  4      6    176.17    4.00                       (----*---)
  5      6    168.43    5.02             (----*---)
                                ----+---------+---------+---------+--
Pooled StDev =   3.81           161.0     168.0     175.0     182.0
```

Based on the p-value = .000 (< .005) there is highly significant evidence to reject the null hypothesis. The difference between the speeds of the five supercars is highly significant.

c   **Tukey's pairwise comparisons**

Family error rate = 0.0500
Individual error rate = 0.00706

Critical value = 4.15

```
Intervals for (column level mean) - (row level mean)

                          1          2          3          4

        2    -23.514
             -10.619

        3     -7.131      9.936
               5.764     22.831

        4    -20.897     -3.831    -20.214
              -8.003      9.064     -7.319

        5    -13.164      3.903    -12.481      1.286
              -0.269     16.797      0.414     14.181

μ₂   μ₄    μ₅    μ₃    μ₁
----------       ----------

        ----------
```

Based on Tukey's test the mean top speeds of a Ferrari and Porsche are the same, of a Lotus and Acura are the same, of a Viper and Lotus are the same. The Ferrari and Porsche are significantly faster than the other three cars. These results are consistent with the appearance of the boxplots.

Unit 4 [Chapters 10-12]

Significant Ideas and Review Exercises

U4.1

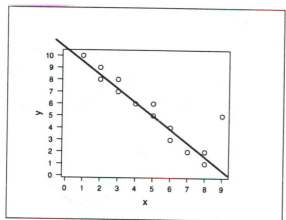

a   Although there is one outlier a straight line should fit the data.
b   The outlier is an influential observation.
d   The influential observation will tend to pull the line toward it resulting in a smaller slope.

U4.3   a   Chi-square test of homogeneity
      b H_0: The reading ability of fourth grade students is the same in the two school districts.
          H_a: The reading ability of fourth grade students is different in the two school districts.
      c   **Chi-Square Test**

```
Expected counts are printed below observed counts

                 C1          C2          C3      Total
       1         21          77          38        136
              28.33       72.64       35.03

       2         34          64          30        128
              26.67       68.36       32.97

Total           55         141          68        264

ChiSq =    1.898 +   0.262 +   0.252 +
           2.017 +   0.279 +   0.267 = 4.975
df = 2, p = 0.084
```

There is mildly significant evidence that the reading ability of fourth grade students is different in the two school districts.

U4.5 a 25% score below 350, 25% between 350 and 425, 25% above 560
   b Chi-square goodness of fit.
   c H_0: $\pi_1 = \pi_2 = \pi_3 = \pi_4 = 1/4$ versus H_a: at least one is not the same
      $\chi^2_{obs} = (4-5)^2/5 + (3-5)^2/5 + (5-5)^2/5 + (7-5)^2/5 = 1.8$; p-value $> .10$
   d The sample does not differ significantly from the national standard.

U4.7 H_0: $\pi_1 = \pi_2 = \pi_3 = \pi_4 = 1/4$ versus H_a: at least one is not the same
      $\chi^2_{obs} = (18-12.5)^2/12.5 + (9-12.5)^2/12.5 + (11-12.5)^2/12.5 + (12-12.5)^2/12.5 = 3.6$; p-value $> .10$
      There is insufficient evidence that one wine is favored over the others.

U4.9 H_0: The mental ages are equally distributed in the intervals: less than 85, 86 to 95, 96 to 110,
       111 to 125, and above 125
       H_a: The mental ages are not equally distributed in the intervals: less than 85, 86 to 95, 96 to 110,
       111 to 125, and above 125
       $\chi^2_{obs} = (3-7)^2/7 + (7-7)^2/7 + (10-7)^2/7 + (8-7)^2/7 + (7-7)^2/7 = 3.71$; p-value $> .10$
       The data support the psychologist's theory; the mental ages are equally distributed in the
       different categories.

U4.11 H_0: The colors are distributed alike for each of the 10-year periods.
      H_a: The colors are not distributed alike for each of the 10-year periods.
      The test is the chi-square test of homogeneity. The p-value ($=.000$) is highly significant.
      Based on these results there is highly significant evidence that color preference has changed over
      the years. The bay and black have gained in popularity.

U4.13 H_0: Political party and opinion on foreign policy are independent.
      H_a: Political party and opinion on foreign policy are dependent.
      Chi-Square Test

      Expected counts are printed below observed counts

	For	Against	Total
1	167	103	270
	136.47	133.53	
2	96	145	241
	121.81	119.19	
3	15	24	39
	19.71	19.29	
Total	278	272	550

      ```
      ChiSq =  6.829 +  6.979 +
               5.471 +  5.591 +
               1.127 +  1.152 = 27.148
      df = 2, p = 0.000
      ```
      There is highly significant evidence that political party and opinion on foreign policy are
      dependent.

U4.15 a   A straight line will fit the data.

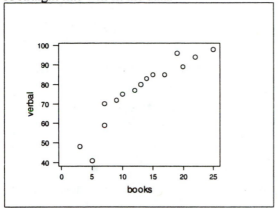

b   The regression equation is   verbal = 45.3 + 2.38 books

c
```
              Fit    Stdev.Fit       90.0% C.I.              90.0% P.I.
For x*=10    69.17       1.94    (   65.74,    72.61)    (   56.83,    81.52)
For x*=15    81.09       1.80    (   77.90,    84.27)    (   68.81,    93.37)
For x*=20    93.00       2.55    (   88.49,    97.52)    (   80.31,   105.69)
For x*=25   104.92       3.68    (   98.40,   111.43)    (   91.39,   118.45)
```

U4.17 a

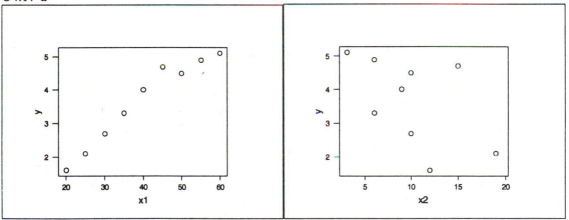

b   The regression equation is $y = -0.508 + 0.0973 x_1 + 0.0271 x_2$

c   R-sq = 95.2% which says that 95.2% of the variability in y is explained by x_1 and x_2.

d H_0: $\beta_1 = \beta_2 = 0$ versus H_a: At least one $\beta_i \neq 0$.
    $F_{obs} = 58.91$ p-value = 0.000   The model is significant.

e
```
Predictor        Coef        Stdev        t-ratio          p
Constant       -0.5079      0.6652        -0.76        0.474
x1              0.09730     0.01071        9.09        0.000
x2              0.02714     0.02993        0.91        0.400
```
Based on the t-ratios x_2 is insignificant. It should be removed from the model. The constant
is also insignificant indicating that the regression line should go through the origin.

U4.19 One-Way Analysis of Variance

```
Analysis of Variance
Source      DF       SS        MS        F        p
Factor       2    135730     67865     2.80    0.069
Error       57   1381410     24235
Total       59   1517140
```

				Individual 95% CIs For Mean Based on Pooled StDev			
Level	N	Mean	StDev				
Major	20	887.0	157.7	(---------*---------)			
Minor	20	904.5	143.8	(---------*---------)			
Private	20	995.5	164.8			(--------*---------)	

```
Pooled StDev =  155.7          840       910       980      1050
```

There is a mildly significant difference between the SAT scores for the different types of universities. Based on the individual confidence intervals it is clear that the difference occurs with the private universities.

U4.21 a

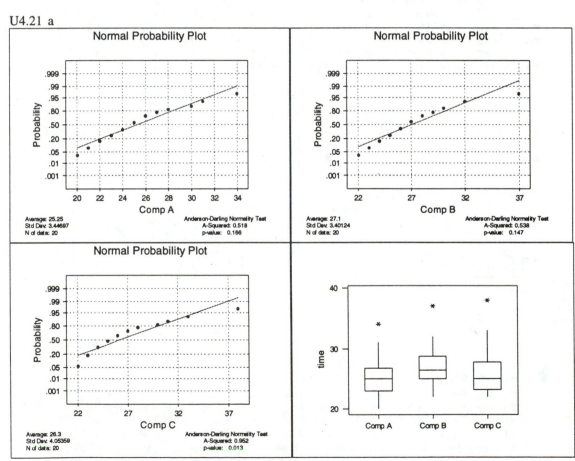

The distributions are slightly skewed right.

b   Because of the robustness of the F-test it could be used in this situation. The skewness is not severe enough to cause problems with the F-test. If, on the other hand, we desired to compare the population medians the Kruskal-Wallis test would also apply.

U4.23 **Chi-Square Test**

Expected counts are printed below observed counts

	Less	6 mos	More	Total
1	192	35	41	268
	150.10	51.51	66.40	
2	124	43	65	232
	129.93	44.59	57.48	
3	135	51	64	250
	140.02	48.05	61.94	
4	174	62	108	344
	192.66	66.11	85.23	
5	121	65	52	238
	133.29	45.74	58.96	
Total	746	256	330	1332

```
ChiSq = 11.699 +  5.290 +  9.714 +
         0.271 +  0.057 +  0.985 +
         0.180 +  0.101 +  0.069 +
         1.807 +  0.256 +  6.086 +
         1.134 +  8.108 +  0.822 = 46.659
df = 8, p = 0.000
```

Based on the chi-square test the duration of time since the last visit to a physician is dependent on the income bracket.

U4.25 a   Normality is a reasonable assumption in all four cases.

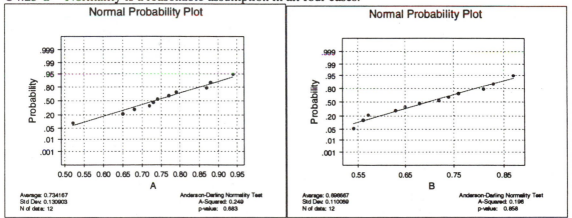

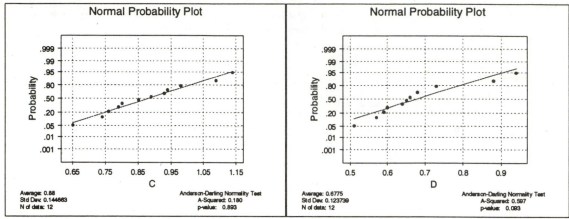

b Based on the boxplots the variances are homogeneous.

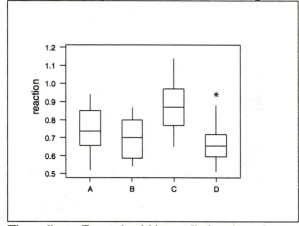

c The ordinary F-test should be applied to these data.

d **One-Way Analysis of Variance**

```
Analysis of Variance
Source      DF        SS        MS        F         p
Factor       3    0.3026    0.1009     6.16     0.001
Error       44    0.7204    0.0164
Total       47    1.0230
                                   Individual 95% CIs For Mean
                                   Based on Pooled StDev
  Level    N      Mean     StDev   ---------+---------+---------+------
A         12    0.7342    0.1309          (------*-------)
B         12    0.6967    0.1101     (-------*------)
C         12    0.8800    0.1447                        (------*------)
D         12    0.6775    0.1237   (-------*------)
                                   ---------+---------+---------+------
Pooled StDev = 0.1280                 0.70      0.80      0.90
```

Based on the p-value there is a highly significant difference between the reaction times of subjects exposed to the four stimuli.